Preface

This handbook aims to provide practical information about bedside glucose testing for nurses, medical technologists, and others who perform or who have management responsibilities for bedside glucose testing. Also, this handbook tries to provide references for more detailed sources of information about specific topics. The intended goal is not to promote either bedside glucose testing or central laboratory services, but to assist in providing the highest quality and most efficient patient care.

Bedside testing is a complex issue that crosses traditional departmental and professional boundaries and that is undergoing rapid evolution due to technological advances, changing regulations, and re-engineering of healthcare delivery and reimbursement. Several issues, such as the effects of bedside testing on costs, quality of test results, and impact on patient care, are highly controversial. This handbook will offer the rationale for opposing viewpoints and the author's conclusions. Others may disagree with some of his conclusions; his biases are based on training as a physician in clinical pathology, 11 years of experience on faculties at academic medical centers, and several years as a laboratory director for the Office of Bedside Testing at the University of Alabama (UAB) Hospitals.

I thank my former coworkers at UAB, including nurses, medical technologists, and physicians such as Carol Howard, William Branum, Luann Hensley, Cherie Gibson, Dana Ellis, Carl Moultrie, Virginia Randolph, Dr. Bruce Alexander, Dr. John Smith, the residents in clinical pathology, and many others; Karla McClellan; colleagues on the Alternative Site Testing Committee of the College of American Pathologists (CAP); Carol Hortin; and many other investigators, speakers, and authors in the field of bedside testing.

Glen L. Hortin, MD, PhD
July 1998

Table of Contents

Part 3. Clinical Care 55

Part 4. Technical Issues 77

HANDBOOK OF Bedside Glucose Testing

by Glen L. Hortin, MD, PhD

Department of Pathology
University of Alabama at Birmingham
Birmingham, AL

Clinical Pathology Department
National Institutes of Health
Bethesda, MD

AACC Press
2101 L Street NW, Suite 202
Washington, DC 20037-1526

www.aacc.org

AACC
Advancing
Clinical Laboratory
Science Worldwide

ISBN **1-890883-05-0**

Printed in the USA

Introduction

1A. Growth of Bedside Glucose Testing

Interest in closely monitoring blood sugar (glucose) levels in patients with diabetes mellitus has skyrocketed over the past decade. Glucose testing represents the highest-volume home-testing activity and, in many healthcare organizations, bedside glucose testing represents the third-highest volume of testing after routine chemistry and hematology.

Many factors have contributed to the expansion in bedside glucose testing over the past decade (see Table 1). The major factor driving this growth is the finding (in clinical trials) that careful control of blood glucose levels can slow the progression of complications from diabetes, for example, kidney disease, vascular disease, and vision loss. Tight regulation of insulin-dependent diabetes, however, requires checking blood glucose levels and dosing with insulin several times daily. Timing of testing, insulin doses, and meals must be coordinated precisely in order to maintain stable glucose levels. Self-monitoring of blood glucose by diabetics using glucose meters has become standard practice and it has led to acceptance of blood glucose meters by physicians, nurses, and patients.

At the same time, significant changes in glucose meter technology have simplified testing procedures and incorporated systems to store test results and control data. Bedside testing has become a more widely practiced testing option as the menu of tests that can be performed has expanded. Glucose testing comprises the highest volume of bedside testing, but bedside testing of blood gases, electrolytes, and other analytes has also grown in volume. Multiple changes in the healthcare delivery system have also increased the demand for bedside testing: cutbacks in phlebotomy and laboratory staffing and consolidation of testing into core laboratories can increase turn-around time; efforts to speed treatment and reduce length of stay emphasize the need for rapid testing and access to results that can be provided by bedside testing. Cost considerations have been both a positive and negative factor—there have been disagreements about whether bedside testing costs more or less than centralized laboratory testing.

Table 1.
Factors Stimulating the Growth of Bedside Glucose Testing

- Discovery that careful monitoring and control of blood glucose levels reduced complications of diabetes
- Increased acceptance of glucose meters by physicians and patients due to growth in home glucose monitoring
- Need for careful coordination and timing of glucose testing, insulin doses, and meals
- Incorporation of glucose monitoring into standards of care
- Improvements in glucose meters
- Downsizing and consolidation of clinical laboratories
- Growth of bedside testing menu and volume
- Cutbacks in phlebotomy services
- Changes in nursing units to "patient-centered" models
- Efforts to reduce patient lengths of stay
- Efforts to reduce costs

Table 2.
Examples of Growth of Bedside Glucose Testing

- Bedside testing programs are relatively new
  A national survey in 1992 found 39% of programs were <3 years old.
- Example of a large teaching hospital
  Lee-Lewandrowski et al describe an increase from 0 tests in 1988 to 60,000 tests in 1992

References

Bickford GR. Decentralized testing in the '90s: a survey of U.S. hospitals. Clin Lab Management Rev 1994(July/August):327-338.

Jones BA, Bachhner P, Howanitz PJ. Bedside glucose monitoring: a college of american pathologists Q-Probes study of the program characteristics and performance in 60 institutions. Arch Pathol Lab Med 1993;117:1080-1087.

Lee-Lewandrowski E, Laposata M, Eschenbach K, et al. Utilization and cost analysis of bedside capillary glucose testing in a large teaching hospital: implications for managing point of care testing. Am J Med 1994;97:222-230.

Rock RC. Why testing is being moved to the site of patient care. MLO 1991;23(9S):2–5.

Lamb LS Jr, Parrish RS, Goran SF, Biel MH. Current nursing practice of point-of-care laboratory diagnostic testing in critical care units. Am J Crit Care 1995;4:429–434.

Kost GJ, Hague C. The current and future status of critical care testing and monitoring. Am J ciln Pathol 1995;104(Suppl 1):S2-17

Thiebe L, Vinci K, Gardner J. Point-of-care testing: improving day-stay services. Nurs Manage 1993;24:56.

1B. Preventing Diabetic Complications by Improved Glucose Control

The key factor in the rapid growth of blood glucose monitoring is the finding that long-term complications of diabetes such as kidney disease, injury to blood vessels, vision loss, and nerve damage depend on how much glucose levels increase. For many years, it was not known whether the many health problems that developed in Type I (insulin-dependent) diabetics were caused by inadequate concentrations of the hormone insulin in the circulation, by the high blood concentrations of glucose that occur in diabetes, or by other unidentified factors. Poor control of blood glucose levels results in increased medical problems and higher costs of care.

A landmark study, the Diabetes Control and Complications Trial, finally provided convincing evidence that many complications of diabetes could be delayed by reducing blood glucose to near-normal levels. This study closely monitored more than 1400 diabetic subjects for an average of nearly 7 years. Major conclusions of the study were described in a report in 1993 and were supported by other studies. However, in order to achieve these lower glucose levels, it was necessary to check blood glucose levels several times a day and to adjust insulin doses according to the blood glucose level. This is a difficult balancing act, because as glucose levels decrease, the risk of serious episodes of hypoglycemia increases. Frequent measurements of blood glucose are critical for regulating glucose levels without causing hypoglycemia.

More recent studies have provided evidence that high glucose levels also contribute to long-term complications in Type II (non-insulin dependent) diabetics. Thus, glucose monitoring is important for this group as well.

Table 3.
Effect of Improved Glucose Control on Diabetes Care

Benefits of Lowering Glucose Levels

- Decreased long-term complications
  - Kidney disease
  - Vision loss due to retinopathy
  - Vascular disease
- Benefits both Type I and Type II diabetics
- Lower costs for medical care

Drawbacks to Lowering Glucose Levels

- Increased risk of hypoglycemia
- More frequent blood glucose testing
- More frequent insulin doses

References

The Diabetes Control and Complications Trial Research Group. The effect of intensive treatment of diabetes on the development and progression of long-term complications in insulin-dependent diabetes mellitus. New Engl J Med 1993;329:977–986.

Reichard P, Nelsson B-Y, Rosenqvist U. The effect of long-term intensified insulin treatment on the development of microvascular complications of diabetes mellitus. New Engl J Med 1993;329:304–309.

Wang PH, Lau J, Chalmers TC. Meta-analysis of effects of intensive blood-glucose control on late complications of type I diabetes. Lancet 1993;341:1306–1309.

Gaster B, Hirsch IB. The effect of improved glycemic control on complications in type II diabetes. Arch Intern Med 1998;158:134–140.

Gilmer TP, Manning WG, O'Connor PJ, Rush WA. The cost to health plans of poor glycemic control. Diabetes Care 1997;20:1847–1853.

1C. How High Glucose Levels Damage Tissues

Glucose reacts chemically with amino groups of proteins at a slow rate that depends directly on glucose concentration. Buildup of damaged proteins can be measured as glycated hemoglobin, one component of which is hemoglobin A1c, or as glycated serum proteins which can be measured as fructosamine. Accumulation of protein damage and further cross-linking and oxidation reactions thicken basement membranes around cells and damage blood vessels, nerves, and kidneys. In accordance with this basic theory of tissue injury in diabetes, it makes sense that decreasing blood glucose levels to near-normal levels will prevent many of the long-term complications of diabetes. The glucose hypothesis—that tissue injury is related to glucose concentrations—together with the evidence from clinical trials that diabetic complications can be prevented by tight glucose control in diabetics provide the rationale for frequent glucose testing and insulin dose adjustment.

References

Brownlee M. Advanced protein glycosylation in diabetes and aging. Annu Rev Med 1996;46:223–234.

Kennedy L, Baynes JW. Nonenzymatic glycosylation and the chronic complications of diabetes: an overview. Diabetologia 1984;26:93–98.

Lyons TJ, Jenkins AJ. Glycation, oxidation, and lipoxidation in the development of the complications of diabetes: a carbonyl stess hypothesis. Diabetes Rev 1997;4:365–391.

Makita Z, Radoff S, Rayfield EJ, Yang Z, Skolnik E, Delaney V, Friedman EA, Cerami A, Valassara H. Advanced glycosylation end products in patients with diabetic nephropathy. New Engl J Med 1991;325:836–842.

1D. Factors Limiting the Growth of Bedside Glucose Testing

Several factors have acted to slow the growth of bedside glucose testing:

1. Federal and state regulations imposed standards and requirements that substantially increase the cost, knowledge of laboratory regulations, and paperwork for a testing program.
2. Competition and turf battles occurred between clinical services that wanted to perform bedside testing and clinical laboratories that wanted to control all testing. A major argument offered by laboratories is that the quality of bedside test results is lower. It is still true that analytical accuracy of glucose meters is lower than measurements by central laboratories although some aspects of performance have improved.
3. Large-scale bedside testing presents a more difficult problem in management than does centralized testing. Performance of testing at large numbers of sites must be tracked and many personnel must be trained.
4. Bedside testing often crosses many departmental boundaries and requires cooperation between groups that usually do not work together closely.
5. There is concern about the inability to capture test results into laboratory or medical data systems.

Some laboratories have responded to demands for faster turn-around time by offering whole blood testing, which can dramatically shorten analysis time. Glucose testing has become an option on an increasing number of models of blood gas and whole blood analyzers. For these analyzers there is no delay for specimen processing and results can be obtained within a time frame similar to blood gas analysis. Following analysis, increased use of bedside data systems for patient care and more effective transmission of laboratory data can improve the speed and efficiency of data transfer to sites where the results are needed.

References

Handorf CR. College of American Pathologists Conference XXVIII on alternate site testing: introduction. Arch Pathol Lab Med (October) 1995;119:867–871.

Watts NB. Bedside monitoring of blood glucose in hospitals: speed vs. precision and accuracy. Arch Pathol Lab Med 1993;117:1078–1079.

Table 4.
Factors Limiting the Growth of Bedside Glucose Testing

- Increased regulation of laboratory testing
- Internal competition for laboratory testing
- Issues of quality of test results
- Increased complexity of managing decentralized testing
- Laboratory whole blood testing to improve central lab turn-around time
- Improved data systems to improve result availability
- Efforts to reduce cost

1E. History of Problems of Glucose Meter Accuracy

Many problems regarding the accuracy of blood glucose meters have been reported. As noted by Greyson, "...complaints about SMBG [self-monitoring blood glucose] devices represent the largest number ever filed with the FDA for any medical device (by the end of 1992, over 3200 incidents, including 16 reports of death, had been filed with the agency)." In the past, problems with glucose meters centered around the critical nature of glucose testing and the importance of assuring testing quality. Improvements in meter technology since that time have reduced but not eliminated the possibility of severe errors.

A major advance has been the introduction of no-wipe test strips that reduce the effect of operator technique on results. Improvements in operator training and quality assurance programs are also significant steps forward. These advances should substantially reduce the number of critical errors and increase detection of errors when they do occur. Some studies have shown evidence for improvement (Figure 5); this was not strongly evident in a national study conducted by the College of American Pathologists.

A large proportion of the severe errors experienced has been attributed to user error. Perhaps it is not surprising that there were severe problems when there was highly variable attention to operator training, equipment maintenance, and quality assurance. Improvements in technology that simplify operation and decrease opportunities for user error should reduce the number of severe errors. Previous experience also underscores the importance of training and quality assurance programs.

References

Greyson, J. Quality control in patient self-monitoring of blood glucose. Diabetes Care 1993;16:1306–1308.

National Steering Committee for Quality Assurance in Capillary Blood Glucose Monitoring. Proposed strategies for reducing user error in capillary blood glucose monitoring. Diabetes Care 1993;16:493–498.

Novis DA, Jones BA. Interinstitutional comparison of bedside blood glucose monitoring program characteristics, accuracy performance, and quality control documentation: a College of American Pathologists Q-Probes study of bedside blood glucose monitoring performed in 226 small hospitals. Arch Pathol Lab Med 1998;122:495–502.

Table 5.
Evidence of Improvement in Meter Accuracy

Comparison in 1994 of new "no-wipe" test strips versus old strips

- No-wipe glucose test systems:
  > 90% of values within 20% of reference method
- "Wipe" glucose test systems:
  80% of values within 20% of reference method

1996 evaluation of meter use for pregnant subjects

- All values within 20% of reference method

References

Li K-L, Huang H-S, Lin J-D, Huang B-Y, Huang M-J, Wang P-W. Comparing self-monitoring blood glucose devices. Lab. Med. 1994;25:585–589.

Stenger P, Allen ME, Lisius L. Accuracy of blood glucose meters in pregnant subjects with diabetes. Diabetes Care 1996;19:268–269.

1F. Tradeoffs in Testing—Speed, Accuracy, and Cost

There are three competing priorities in most laboratory testing—speed, accuracy, and cost. Efforts to improve one usually result in a tradeoff: another priority may be superceded. The traditional view as described by Handorf is:

"Good quality, low cost, or fast turn-around time … you can have any two."

These constraints apply both to central laboratory and to bedside testing. Laboratories do not strive for the highest possible accuracy, because the cost would be prohibitive and the turn-around time too slow for them to apply the most accurate methods (such as isotope-dilution mass spectrometry) to routine specimens. There is a balancing act involved when trying to meet clinical needs for accuracy and speed at reasonable cost. Efforts to improve turn-around time in a laboratory often result in substantial costs due to increased staffing or purchase of additional equipment.

The traditional view is that bedside glucose testing is fast and cheap but results are poor. The record of problems, many of them related to poor operator performance and lack of quality assurance, have produced this impression, particularly among laboratory workers who focus on analytical quality rather than the total process of diabetes management. Also, misunderstandings about differences of measurement in glucose concentration of whole capillary blood vs. that in venous plasma contribute to this conclusion

Systematic factors that might affect comparative studies with laboratory methods are described in Section 4, Technical Issues. Advances in technology and quality assurance that explain how the foregoing record of glucose meter performance has improved appear in Section 5, Quality Assurance Issues. Further improvement in accuracy is still a desirable goal.

Table 6.
Tradeoffs in Testing

Tradeoffs among major priorities—efforts to improve one usually worsen another

- Accuracy
- Speed
- Cost

Historical performance of bedside glucose meters

- Low accuracy
- Good speed
- Low cost

Goals for improving meter performance

- Improve accuracy
- Do not sacrifice speed or cost

References

Handorf CR. College of American Pathologists Conference XXVIII on alternate site testing introduction. Arch Pathol Lab Med (October) 1995;119:867–871.

National Steering Committee for Quality Assurance in Capillary Blood Glucose Monitoring. Proposed strategies for reducing user error in capillary blood glucose monitoring. Diabetes Care 1993;16:493–498.

Watts NB. Bedside monitoring of blood glucose in hospitals: speed vs. precision and accuracy. Arch Pathol Lab Med 1993;117:1078–1079.

1G. Changing Technologies for Glucose Measurement

There have been rapid and dramatic changes in the technology applied to monitoring glucose levels in diabetics. The first approach used was to detect glucose excreted in urine. The limitation of this approach is that there is usually little excretion of glucose in urine unless the blood glucose concentration exceeds 200 mg/dL. Thus, this approach serves as an indicator of hyperglycemia, but it does not help monitor blood glucose levels within currently applied target ranges and it does not indicate whether blood glucose levels are getting too low.

The technology for blood glucose testing has undergone several major evolutions. First, there were manual test strips that required blotting and visual interpretation against a color chart. Next, meters were developed to provide more precise color readings. These required manual timing and wiping of the strip before inserting it in the meter for reading. The manual timing and blotting were weak links in performance: test results differed depending on how hard the strips were wiped or blotted. Depending on their technique, different meter operators got different results. Finally, "no-wipe" strips that decrease the technique-dependent variability of measurements were developed.

There have been other important technological developments. Systems have been added to store data on patient results, operators, and quality control. Meters are better able to detect malfunctions or inadequate specimens. Control of manufacturing processes has led to improved reliability of test strips and meters. The growing demand for glucose measurements has accelerated the pace of change, and future technologies may include less invasive methods of sample collection or totally noninvasive glucose measurement by external sensors.

Reference

Chmielewski SA. Advances and strategies for glucose monitoring. Am J Clin Pathol 1995;104(Suppl 1):S59–S71.

Table 7.
Changes in Glucose Testing Technology

Urine glucose testing

- Manual urine dipsticks
- Only detects marked hyperglycemia, glucose > 200 mg/dL

First-generation blood testing: manual

- Visual estimate from test strip
- Strips require wiping

Second generation: reflectance meters

- Manual timing and wiping
- Glucose oxidase chemistry
- Addition of data systems

Third-generation: no-wipe meters

- Elimination of wiping step
- New measurement principles
  - Electrodes
  - New chemistries
- Improved error detection
- Enhanced data systems

1H. Improvements in Meter Performance

Technological changes have improved meter performance. In this author's view, the most important improvements have been those that make consistent results independent of meter operator. It is important that many operators with widely varying levels of skill and training all produce equivalent results. The wiping of test strips was previously the step with greatest potential for operator variability; eliminating this step helps achieve more consistent results regardless of operator. Any testing programs that use older generations of strips, which require wiping before reading, should change to decrease this variable.

The measuring range of glucose meters has been extended so that very low or very high glucose levels can be measured. There has been a general trend toward reducing the volume of specimen required and increasing the ability of meters to detect if inadequate volume of specimen was provided. These are important steps in decreasing errors from inadequate specimen size. Improved internal diagnostics check whether batteries, electronics, and optical paths are working and help identify when meters are broken or require cleaning. A one-year survey conducted at a large medical center found that more meter problems were identified by internal diagnostics than by routine quality control measurements. Other performance improvements include the ability to measure other sample types such as venous blood; to detect errors in specimen addition or meter operation; and, by addition of computers to meters, to track patient data, operators, and quality control data.

These performance improvements should improve accuracy, but there is little proof that this has occurred. It is likely that most improvement in accuracy will occur as a result of eliminating a small number of critical blunders, including mistakes by operators with the least skill and experience who only occasionally perform testing. And glucose testing reference methods, which display systematic biases depending on type of sample used (venous or capillary), complicate comparison studies. These complications and comparisons are discussed later in this section and in Section 4, Technical Issues.

Reference

Averyt JM, Howard C, Hortin GL. Detection of problems with glucose meters [abstract]. Clim Chem 1997;43:S140.

Table 8.
Improvements in Meter Performance

Results of technological advances

- Decreased dependence on operator technique
- Improved precision of measurement
- Simpler operation
- Wider glucose measuring range
- Decreased specimen volume
- Acceptance of venous blood samples
- Shorter analysis times
- Improved detection of meter malfunctions
- Improved detection of inadequate specimens
- Decreased interferences such as altitude effects, effects of blood oxygen concentration, effects of reducing compounds
- Improved data handling
- Improved quality assurance of testing
- Decreased maintenance required

References

Chmielewski SA. Advances and strategies for glucose monitoring. Am J Clin Pathol 1995;104(Suppl 1):S59–S71.

A survey of features of bedside glucose testing systems is provided in CAP Today 1997(June.);10:33–39.

Chmielewski S, et al. Precision and accuracy of the Accu-Chek Advantage blood glucose monitoring system at high altitude. Clin Chem 1996;42:115–117.

Giordano BP, et al. Performance of seven blood glucose testing systems at high altitude. Diabetes Educator 1989;15:444-448.

1I. Do Glucose Meters Meet Clinical Accuracy Goals?

Most glucose meters do not meet accuracy goals for applications such as screening or diagnosis of diabetes or measurement of glucose from glucose tolerance tests. Most meters are not suitable for these applications.

For the primary application of meters—the monitoring of glucose levels in diabetics—varying accuracy goals for clinical use have been set. A common standard is that glucose meters should yield values within 20% of "true" values ascertained by a reference method. Clinical accuracy goals depend to some degree on the level of glucose being measured. A common way of classifying errors of measurement has been Clarke's error grid analysis. Standards for accuracy and precision of glucose meters proposed by the American Diabetes Association to meet clinical needs have become progressively more stringent as technology has improved.

A study of physician goals and meter performance in 1994 concluded that meters provided adequate measurements at normal and high glucose levels but provided inadequate measurements at glucose levels <85 mg/dL. Standard deviations of measurements were about 7% at 150 mg/dL and 14% at 85 mg/dL.

References

Clarke WI, Cox D, Gonder-Frederick LA, Carter W, Pohl SL. Evaluating clinical accuracy of systems for self-monitoring of blood glucose. Diabetes Care 1987;10:622–628.

Davis M, Walker EA. Capillary blood glucose monitoring for clinical decision making. Lab Med 1992; 23:591–595.

Watts NB. Reproducibility (precision) in alternate site testing: a clinician's perspective. Arch Pathol Lab Med 1995;119:914–917.

Weiss SL, Cembrowski GS, Mazze, RS. Patient and physician analytic goals for self-monitoring blood glucose instruments. Am J Clin Pathol 1994;102:611–615.

Table 9.
Performance Goals for Glucose Meters

- 1986, American Diabetes Association: Precision goal ±10% (coefficient of variation)
- 1987: Goal for total error < 20%
- 1991: 90% of laboratories achieved precision better than ±10% (Standard deviation)
- 1996: American Diabetes Association: "The goal of SMBG device manufacturers should be to make future SMBG systems with an analytic error ±5%."

SMBG: self-monitoring of blood glucose

References

American Diabetes Association: Consensus statement on self-monitoring of blood glucose. Diabetes Care 1987;10:93–99.

American Diabetes Association. Self-monitoring of blood glucose. Diabetes Care 1996;19:S62–S66.

Clarke WI, Cox D, Gonder-Frederick LA, Carter W, Pohl SL. Evaluating clinical accuracy of systems for self-monitoring of blood glucose. Diabetes Care 1987;10;622–628.

Jones BA, Bachner P., Howanitz PJ. Bedside glucose monitoring: a College of American Pathologist Q-Probes study of the program characteristics and performance in 605 institutions. Arch Pathol Lab Med 1993;117:1080–1087.

Management Issues

2A. Do You Need Bedside Glucose Testing?

The first and one of the most important management issues with respect to bedside glucose testing is whether it is needed at all in your care setting. All of the advantages and disadvantages need to be weighed. The need increases if turnaround time for glucose test results from laboratories is long, if there are many stat glucose tests and diabetic patients, and if there is a strong need to conserve on the volume of blood drawn (usually pediatric testing). There is less need if a laboratory can provide rapid results, if there are few stat blood glucose levels needed, and if the need for blood volume conservation is less acute—for example, in patients who are all adults with little demand for transfusion. When there is a high volume of glucose tests requiring rapid results, as for management of diabetics, bedside testing becomes more attractive because costs decrease (see Cost Issues, Section 2M), and laboratories struggle to provide rapid results consistently on many tests.

The speed and ease of blood specimen transport can be a major factor in whether bedside glucose testing is needed. Hand transport of specimens from the bedside to the laboratory is labor intensive and can be a major source of delays. Investments in infrastructure (such as installing a pneumatic tube system) might be an alternative to bedside testing. Similarly, improvements in the communication of test results from the laboratory may be a key factor in meeting turn-around need. Rapid analysis offers little benefit if results cannot rapidly reach the point of use.

Table 10.
Assessing Need for Bedside Glucose Testing

High Need	Low Need
Slow lab glucose results	Fast lab results Fast sample transport Fast analysis Fast reporting
Many glucose tests	Few glucose tests
Many diabetics	Few diabetics
High need to conserve blood Newborn patients Transfusion-dependent patients	Low need to conserve blood Adult patients Few transfusions
Limited phlebotomy service	Extensive phlebotomy service

2B. The Need for Rapid Glucose Results for Diabetes Care

Most glucose tests for diabetes care are pre-ordered and scheduled a day or more in advance. Usually, if a test is ordered in advance, there is no great urgency to obtain quick results, and laboratory workers may not recognize a high priority for rapid results. However, the optimal turn-around for glucose testing for diabetes management is similar to what is commonly provided by laboratories for urgent or stat testing.

There are two primary reasons why rapid turn-around time for glucose testing is desirable. First, physiological changes in blood glucose levels can occur quite rapidly. Levels can change drastically within 15–30 minutes of food ingestion, stress, or insulin doses. Drawing specimens for glucose monitoring two hours in advance may allow a patient to become hypoglycemic or hyperglycemic before any adjustment in care is made. Also, waking patients at 5 a.m. to draw their blood does not contribute to their satisfaction and normal physiological schedule. Changes in sleep patterns or increased stress may affect levels of cortisol and adrenaline that in turn affect efforts to control blood glucose.

The second reason is that tight control of glucose levels requires careful and consistent coordination of glucose testing, insulin doses, and meals. Short-acting insulin (usually included in doses before meals) should be administered about 30 minutes before meals. Delays in obtaining test results require delaying insulin doses and then delaying meals. Disrupting the daily schedule will change the intervals between meals, altering the daily pattern of glucose control; efforts to regulate glucose levels may be hindered. This can be a very costly problem if it translates into additional days in the hospital to achieve stable glucose levels. Also, nursing unit and meal delivery schedules may be disrupted, resulting in lower efficiency. Unpredictable turn-around times for glucose results make it difficult to schedule patient care.

Table 11.
Why Fast Glucose Results Are Needed for Care of Diabetics

- Physiological changes in blood glucose occur rapidly.
  Glucose levels may change before test results come back.
- Glucose testing, insulin doses, and meals must be coordinated. Delays in glucose test results affect timing of insulin and meals.
- Consequences of slow glucose test turn-around time
  - Increased chance of hypoglycemic episodes
  - Need to wake patient early for blood draw
  - Disruption of timing of insulin doses and meals, resulting in more variable glucose control
  - Decreased efficiency of nursing units

2C. Inventory of Testing Volumes

One of the most basic elements in assessing the need for bedside glucose testing is to determine the volumes of testing currently performed at the bedside and in the central laboratory. This may be surprisingly difficult, and it can serve as an example of why there should be improved management of bedside testing. You cannot adequately plan or manage activities that you cannot measure. In many cases, it has been a challenge to measure the volume of bedside glucose testing when it has been performed independently by multiple nursing units or departments. For example, nursing units may be poor in responding to surveys sent out by laboratory staff because the survey may be viewed as an effort to control their activities.

Estimates of testing volumes may be obtained from purchasing departments or vendors by compiling the number of meters and test strips purchased. However, this may not accurately estimate patient testing if there are no data indicating how many test strips were used for quality control testing or training or how many were sent home with patients or wasted. In some large organizations it is hard even to identify all of the areas that perform testing. Information about laboratory licenses and about test supply purchases may help identify all testing sites.

Assessing the number of central laboratory tests that will be replaced by introducing glucose meters is difficult because not all glucose measurements are ordered as individual tests. Estimates may need to be based on the number of diabetic patients admitted and the expected volume of glucose testing alone. If critical pathways have been developed, these may identify the typical number of glucose tests for the usual diabetic care and may allow you to estimate the number of tests that will be performed at the bedside.

Table 12.
Inventory of Glucose Testing

Inventorying bedside testing can be a problem.

- Surveys of nursing units

  (But responses may be incomplete and data on test volume lacking)
- Purchasing departments may know purchasing volume

  (But data on patient usage may be incomplete—how many for patients, QC, discarded, sent home with patients, etc.)
- Vendors may provide volume of purchases.

  (But no data on patient usage)
- Laboratory licenses may identify sites performing tests.

  (But little information on test volumes)

Central laboratory glucose testing

- Number of glucose tests ordered individually

  (But how many would be replaced by bedside testing?)
- Glucose ordered as part of panels

  (Would any be replaced by bedside testing?)

2D. Alternatives to Bedside Glucose Testing

Before you decide whether bedside glucose testing is needed, evaluate the alternatives. If your laboratory can consistently report glucose test results within 30–60 minutes of sample collection, it may be possible to use this as an alternative to bedside testing. This would require a within-lab turn-around time of about 15–30 minutes to allow time for sample transport and reporting of results. Rapid analysis within a laboratory requires either whole blood analysis or highly expedited processing and analysis. Collection in gray-topped or heparinized blood collection tubes is necessary because there is not sufficient time for blood to finish clotting to generate serum.

Improving the speed of laboratory analysis may be an efficient alternative to bedside glucose testing when the volume of glucose tests ordered as individual tests is low, and when the laboratory is already set up to deliver rapid turn-around time for other tests, such as blood gases. However, there are several major hurdles that need to be overcome in providing rapid within-laboratory analysis (see Table 13). Key success factors are communication with clinical services that need glucose testing, agreement on definitions and goals for turn-around time, and the ability to work across boundaries within an organization. These may be more difficult than the technical issues of how to collect, process, and analyze specimens.

Then, too, there is the recurring challenge of how to do more, faster, and cheaper. It is possible to improve turn-around time by hiring more staff, buying more equipment, and setting up more satellite laboratories, but any option that substantially increases costs is rarely an acceptable solution in the current environment of cost-containment. A final challenge involves developing a monitoring system so that you can review whether goals are being met and where problems are occurring.

References

Green M. Successful alternatives to alternate site testing: use of a pneumatic tube system to the central laboratory. Arch Path Lab Med 1995;119:943–949.

Fleisher M, Schwartz MK. Automated approaches to rapid-response testing: a comparative evaluation of point-of-care and centralized laboratory testing. Am J Clin Pathol 1995;104(Sullp 1):518–525.

Felder RA, Savory J, Margrey KS, Holman JW Boyd JC. Development of a robotic near patient testing laboratory. Arch Pathol Lab Med 1995;119:948–951.

Table 13.
Alternatives to Bedside Glucose Testing

Challenges that need to be overcome

- Agreeing on definition and expectation of turn-around time
- Working across departmental boundaries
- Rapid transport of specimens to the laboratory
- Rapid analysis of blood glucose
    - Collection in gray- or green-topped tubes
    - Whole blood analysis or expedited processing and analysis of plasma
- Expedited transmission of results to clinical staff
- Keeping cost of testing low
- Developing a monitoring system

2E. Rapid Transport of Specimens

Specimens transport can be a major source of delay if the laboratory is distant from patients. If the laboratory is more than a two- to three-minute walk from blood-drawing sites, use of mechanical transport systems will probably be desirable. These may include dumbwaiters, conveyors, or pneumatic tube systems. Mobile robots have sometimes been used to transport specimens in hospitals, but their speed is usually slower than that of a person walking. Setting the laboratory in a central location near patient care areas and providing sample transport via pneumatic tube may solve transport issues. An alternative strategy is to site satellite laboratories near patient care areas. This solves the transport problem for selected areas, but usually increases expenses for facilities, equipment, and staffing. Recent developments in analyzers, simplification of quality control procedures, and use of computer networks permit a central laboratory to perform many supervision and management functions remotely. This offers some opportunities to develop alternative models of satellite laboratories in which testing is performed by nursing rather than by laboratory staff.

Table 14.
Potential Solutions to Transport Delays

- Mechanical transport systems
  - Pneumatic tubes
  - Dumbwaiters
  - Conveyors
  - Mobile robots (usually too slow for stats)
- Satellite laboratory near patients
  - Staffed by laboratory workers
  - Testing by nursing staff, management by laboratory

Green M. Successful alternatives to alternate site testing: use of a pneumatic tube system to the central laboratory. Arch Path Lab Med 1995;119:943–949.

Fleisher M, Schwartz MK. Automated approaches to rapid-response testing: a comparative evaluation of point-of-care and centralized laboratory testing. Am J Clin Pathol 1995;104(Sullp 1):518–525.

Felder RA, Savory J, Margrey KS, Holman JW Boyd JC. Development of a robotic near patient testing laboratory. Arch Pathol Lab Med 1995;119:948–951.

Howanitz PJ, Sunseri DA, Love LA, Lohr A. Adapting mobile robotic technology to intralaboratory specimen transport. Arch Pathol Lab Med 1996;120:944–950.

2F. Expectations for Laboratory Turn-Around Time

Physicians commonly expect results of a stat glucose test to be available within 15–30 minutes, whereas laboratory professionals often consider a turn-around time of <60 minutes good performance. To maintain tight control of diabetes, test results should be known within 60 minutes to avoid delays in meals and medication. For critical samples, the turn-around time should be <30 minutes.

Clinicians and laboratory staff need to agree on turn-around time goals because these goals will determine needs for instrumentation, staffing, and other aspects of laboratory organization. Collection of data about turn-around times is useful to prevent deterioration of service, to troubleshoot sources of problems, and to assure clinicians about performance. Data collection on turn-around time can be an important factor in gaining physician acceptance of laboratory testing, because without data on overall performance, there is a tendency to focus on and magnify the significance of a small number of failures.

References

Jahn M. Turn-around time down sharply, yet clients want results faster. MLO (Sept) 1993;:24–30.

Jones BA, Bachhner P, Howanitz PJ. Bedside glucose monitoring: a college of american pathologists Q-Probes study of the program characteristics, and performance in 605 institutions. Arch Pathol Lab Med 1993;117:1080–1087.

Hilborne LH, Oye RK, McArdle JE, Repinski JA, Rodgerson DO. Use of specimen turn-around time as a component of laboratory quality: a comparison of clinician expectations with laboratory performance. Am J Clin Pathol 1989;92:613–618.

Table 15.

Turn-Around Time Expectations for STAT Glucose Tests

- Differing expectations
    - Physicians usually expect <30 minutes (from collection to receipt of results)
    - Laboratory workers usually expect <60 minutes (from receipt in lab to reporting)
- Clinical staff will probably feel a need for bedside glucose testing if
    - Turn-around time for STAT tests is > 30 minutes
    - Turn-around time for testing of diabetics > 60 minutes
- Try to agree on goals
- Develop a monitoring system

2G. Defining Laboratory Turn-Around Time Goals

A major source of conflict between clinical and laboratory staffs is their different views of turn-around time. Physicians and nurses usually view turn-around time as the time from placing an order or drawing a sample until they receive the result. Laboratorians tend to view turn-around time as the time from receipt of sample to entry of results into a laboratory computer system or transmission of a report. Laboratory workers are reluctant to take responsibility for total process time because they feel they have little control over steps occurring outside the laboratory. Also, it may be difficult to track draw times and delivery times unless there is very good compliance with recording of specimen draw times.

Lack of agreement on how turn-around time is measured will cause ongoing conflict between the laboratory and the nursing unit. Technologists in the laboratory will think they are doing an outstanding job, and they will not understand why physicians and nurses are upset about, what is from the clinical perspective, poor turn-around time.

Before a goal for turn-around times can be set, the parameters for it must be clearly defined. From a clinical care standpoint, the important interval for glucose testing is the interval from specimen collection to the time that a nurse receives the result. (Most glucose testing for diabetes is pre-ordered so that the interval from order to specimen collection is not important.) Meeting turn-around time goals using this definition will require efficient sample transport, fast analysis, and effective transmission of results to caregivers.

Table 16.
Defining Laboratory Turn-Around Time

Differing viewpoints

- Physician's or nurse's view: from order or collection to receipt of result
- Laboratory's view: from sample receipt in lab to sending report

Need to agree on how turn-around time is defined

- Clinically useful interval for scheduled glucose testing: from sample collection to nurse receipt of result

2H. Benefits and Shortcomings of Bedside Glucose Testing

Bedside glucose testing is often viewed simplistically as a tradeoff between faster results and lower analytical accuracy. These are two of the most important issues, but there are many other factors that determine whether bedside glucose testing will be beneficial or detrimental in a particular healthcare setting. The decision to adopt bedside glucose testing should be based ultimately on what will deliver the highest-quality and most-efficient patient care.

Some benefits of bedside glucose testing include being able to perform testing on capillary specimens and reducing the volume of blood drawn from about 3–10 mL for a typical venous specimen to about 0.1 mL. Another benefit involves significant gains in testing quality: for example, testing a fresh specimen so that glucose levels do not change during specimen transport; being able to directly compare test results with clinical status and treatments; being able to repeat results immediately; avoiding sample mislabeling or mix-up; and decreasing communication errors that may occur via telephone reports or verbal communication among staff.

Other benefits impact cost and patient care. Bedside testing may improve patient care by avoiding hypoglycemic episodes that could occur when test results and meals are delayed, and it can improve glucose control by keeping insulin doses and meals on a regular schedule. Nursing units can increase efficiency by being able to schedule care predictably. Bedside glucose testing can also improve the efficiency of central laboratory stat testing: by eliminating the need to do glucose testing (which, due to volume, can slow other test results processing) results can be reported more quickly. Patient satisfaction and education can be promoted by letting them sleep longer than they could if they had to be awakened for phlebotomy and by demonstrating to them how glucose testing should be performed when they go home. Costs can also decrease (if testing volume is high); this issue is discussed in "Cost Issues," section 2M.

Besides the major shortcoming of lower analytical accuracy, bedside glucose meter testing has other limitations. Performance of testing at a large number of sites requires more time for training and quality assurance, and decentralized testing is harder to manage. There is likely to be a significant volume of duplicate testing (specimens may be tested for glucose both at the bedside and in the central laboratory). It is harder to capture bedside testing results into laboratory or medical information systems, so it may be more difficult to find previous days' results and to track glucose control over periods of time. And if results are not entered in information systems, it is difficult to bill for tests or to track patterns of utilization.

Table 17.
Benefits and Shortcomings of Bedside Glucose Testing

Benefits of Bedside Testing	Shortcomings
Faster test results	Less accurate results
Decreased blood volume drawn	Results may not reach laboratory record
Use of capillary samples	Harder to manage testing
No need to delay meals and medication	More time for training
Fresh sample	More time for quality assurance
More rapid detection of hypoglycemia	Testing by non-laboratorians
Fewer process steps	More duplicate testing
Comparison of test results vs. patient condition	Harder to find previous test results
Avoid overload of lab with stats	Billing is more difficult
Avoid mislabeling of specimens	
Immediate repeats when needed	
Patient education and comfort (no need to wake patient as early for phlebotomy)	

Table 18.
Goals of Glucose Testing

- To improve linkage of test results to patient care
- To improve patient outcomes
- To increase efficiency of care
- To minimize costs
- To reduce blood losses from phlebotomy
- To improve patient satisfaction
- To promote patient education

Which option—bedside or central laboratory testing—best meets these goals for your care setting?

2I. Starting a Centralized Bedside Glucose Testing Program

If bedside glucose testing is occurring in many patient care units, there may be a number of significant advantages to centralizing testing management. Centralized management generally reduces labor for quality assurance and training activities and for meeting regulatory requirements. Standardization of equipment and reagents often saves money. If it is not possible to completely centralize operations into a single management structure, there should at least be coordination and exchange of information between different testing sites. Often there is little action until a crisis is recognized, such as possible loss of accreditation, poor evaluations on inspections, inspection imminent but operations unprepared, physician complaints, inability to provide adequate laboratory service, or bad patient outcomes. Ideally, planning should address not only the immediate crisis but use change as an opportunity for long-term improvement.

Whoever initiates the effort to develop a centralized program must recognize the political need to identify and gain support from the key leadership, who will have to approve the program. In different organizations, this may include hospital administrators, nursing directors, laboratory directors, physicians providing diabetes care, laboratory administrators, or others whose budget, personnel, or patient care will be affected.

A committee should be organized to represent and communicate with the key departments and to help secure their buy-in. Failure to obtain support from even one key department may serve as an insurmountable roadblock to progress or may undermine success once a program begins.

References

Kurec AS. Implementing point-of-care testing. Clin Lab Sci 1993;6:225–227.

Lewandrowski K, Cheek R, Nathan DM, et al. Implementation of capillary blood glucose monitoring in a teaching hospital and determination of program requirements to maintain quality testing. Am J Med 1992;93:419–426.

Harris CH, Utz C, Gibson C. A pain-free inauguration of point-of-care testing. MLO June 1995;27:41–43.

Table 19.
Steps in Organizing a Centralized Bedside Testing Program

- Identify a reason for change
  - Response to a recognized problem or crisis
    - Impending inspection for accreditation
    - Identified deficiencies in bedside testing
    - Physician complaints
    - Poor laboratory turn-around time
    - Bad patient outcomes
  - Part of re-engineering efforts
    - Laboratory downsizing
    - Changes in diabetes care processes
  - Cost reduction efforts
- Obtain leadership support of affected areas
  - All areas where testing will affect budgets, personnel, or patient care
- Form a committee to plan change
  - Represent all areas
    - Teamwork
    - Effective communication
    - Diverse knowledge base

2J. Organizing a Bedside Glucose Testing Program

In order to sell all affected departments on the importance of centralized management of bedside testing, it is important to identify which institutional problems will be solved and what added value will be traded for loss of autonomy. Some ways in which improved management might provide value are listed in Table 20. The importance of different elements may vary for specific organizations.

The committee process usually offers slow results, but for efforts like bedside testing that cross several organizational boundaries, it is important that all key departments participate in decision making. This is necessary in order to take advantage of the specific knowledge and skills of different groups as well as to open up communication and break down barriers between departments. It is important to develop teamwork, because bedside testing usually needs both laboratory practice and patient care skills.

Organizing a program is a long-term process, not a quick fix for an impending inspection. Planning should begin a year or more before anticipated inspections. An established track record of performance is needed for inspections to be favorable.

Involve all key departments in planning and decision making. Bedside testing is a process that crosses many areas of responsibility, and setting up a program is a big job. Spread the work out, and start developing teamwork.

Partnership between nursing, laboratory, and medical staff is important for the success of bedside testing programs, because different groups can contribute useful skills based on their individual expertise. For a description of how to promote teamwork in starting a bedside testing program, consult the references listed below.

References

Harris CH, Utz C, Gibson C. A pain-free inauguration of point-of-care testing. MLO June 1995;27:41–43.

Ingram-Main R, Kiechle FL. Implementing a successful bedside glucose program. MLO (April) 1993;25:25–28.

Kurec AS. Implementing point-of-care testing. Clin Lab Sci 1993;6:225–227.

Table 20.
Value of Centralized Management of Bedside Testing Programs

Potential components of added value

- Improved patient care
- Improved test quality
- Cost reduction
- Decreased blood losses from phlebotomy
- Improved patient satisfaction
- Improved worker satisfaction
- Tracking of utilization
- Approval by regulatory agencies
- Improved risk management
- Improved test data retrieval

Table 21.
Key Groups Needing Input in Planning Bedside Glucose Testing

- Nursing
- Laboratory administration
- Hospital administration
- Diabetes educators
- Laboratory directors/pathologists
- Physicians providing diabetes care
- Goals of the planning group
  - Evaluate needs for bedside testing
  - Assess needed and available resources
  - Develop a management structure

2K. Nursing and Laboratory Inputs to Bedside Testing

Contributions of nursing and laboratory services are complementary. Nursing staff provide expertise in clinical care and nursing standards. Laboratory staff contribute knowledge of laboratory standards, regulations, test equipment and reagents, and quality control procedures. Bedside testing requires a wide mix of skills and knowledge; effective partnership between nursing and laboratory workers is often critical for success. Teamwork can be promoted by establishing an atmosphere of mutual respect for the specific skills and contributions nursing and laboratory staff bring to the testing task. Recent increased regulation and attention to laboratory standards for bedside testing make laboratory staff assistance in meeting these standards particularly valuable.

Working with bedside testing can be challenging for laboratory technologists who are used to the more controlled and structured environment of the clinical laboratory, but it also offers an opportunity to contribute their skills to new testing applications. For nursing staff, the major challenge is in acquiring an appreciation for routine quality assurance procedures for laboratory testing; their major focus has traditionally been on patient care issues.

References

Roby PV, Kenny MA, Garza D. The laboratory outside the laboratory: our role in point-of-care testing. Clin Lab Sci 1993;6:222–225.

Yablonsky T. Point-of-care testing: the evolving role of the medical technologist. Lab Med (December) 1994;25:777–780.

Table 22.
Complementary Knowledge of Nursing and Laboratory Staffs

Nursing	**Laboratory**
Patient care standards	Quality control practices
Use of test data	Test procedures and manuals
JCAHO standards	CLIA license acquisition
Diabetes education	CAP standards

2L. Organizational Issues for Bedside Testing

There are a number of difficult organizational issues that should be addressed when a bedside glucose testing program is set up. For instance, the responsibilities of different personnel should be defined and assigned; quality assurance and oversight of regulatory issues, for example, are often assigned to a person with a laboratory background. If a laboratory staff member or pathologist is made the director of bedside testing, that person must create a close working relationship with the nursing director. If a nursing staff member is made director, the laboratory director must effectively support the nursing staff member. Staff who perform the tests should be accountable to whomever provides their performance evaluations; this provides the most direct and effective feedback to good or poor performance.

There are a variety of ways to handle budgets for a bedside testing program. Separating out all test coordinator, equipment, and supply costs into a central bedside testing budget allows better tracking, but it has the drawback of lower accountability and less incentive to avoid waste. With respect to revenue, decide how monies from testing should be allocated.

Because of the risk to patients and the accountability of the program director to regulatory agencies, bedside testing program personnel should have the authority to remove meters from units that do not meet acceptable standards. Establish clear standards and performance criteria to avoid arbitrary restriction of services.

Federal regulations indicate only that testing personnel should be high school graduates and should receive adequate training. Some state regulations are much more restrictive about who can perform laboratory testing. Potential candidates for performing testing are nurses, patient care technicians, nursing aids, phlebotomists, and laboratory technologists. Many factors come into play in deciding who has the time and skills to perform testing.

Decisions about how many CLIA licenses should be obtained and which accrediting agency will be used need to be made. Many testing sites at a single address can be covered by a single CLIA license, but laboratories may be reluctant to combine with units for which they do not have control over quality.

Reference

Lamb LS. Responsibilities in point-of-care testing: an institutional perspective. Arch Pathol Lab Med 1995;119:886–889.

Table 23.
Organizational Issues for Bedside Glucose Testing

- Who is the director of bedside testing?
- To whom do test coordinators report —nursing, laboratories, or pathology?
- What department budget is responsible for bedside testing?
- Do personnel have authority to remove meters from poorly performing or low-volume units?
- Who performs testing?
  - Considerations:
    - State regulations
    - Current testing personnel
    - Available staff
    - Union contracts (if unionized)
    - Skills and training issues
- How is revenue allocated?
- How many CLIA licenses will be obtained?
- What accreditation will be sought?

2M. Cost Issues

Views about the costs of bedside glucose testing relative to glucose analysis by a central laboratory vary widely. Many laboratory workers hold dogmatic opinions reflective of those expressed by Winkelman: "The costs of central laboratory testing are always very much less than that of distributed testing." Others promote bedside testing as an important mechanism for cost reduction, citing savings in direct costs of performing tests or as difficult-to-measure savings from improved patient care due to faster turn-around time. Workers often believe that faster receipt of test results should improve patient outcomes and nursing unit efficiency, but it has been difficult to measure improvements. Getting a better handle on total program costs of bedside glucose testing clearly is a necessary step in evaluating cost effectiveness; different cost estimates may lead to different conclusions about the most cost-effective testing delivery method. Cost analysis has been one of the most complicated and controversial issues related to performance of glucose or other tests at the bedside.

References

Greendyke RM. Cost analysis: bedside glucose testing. Am J Clin Pathol 1992;97:106–107.

Winkelman JW, Wybenga DR, Tanasijevic MJ. The fiscal consequences of central vs. distributed testing of glucose. Clin Chem 1994;1628–1630.

Nosanchuk JS, Keefner R. Cost analysis of point-of-care laboratory testing in a community hospital. Am J Clin Pathol 1995;103:240–243.

Statland BE, Brzys K. Evaluating stat testing alternatives by calculating annual laboratory costs. Chest 1990;97:198S–203S.

Lee-Lewandrowski E, Laposata M, Eschenbach K, et al. Utilization and cost analysis of bedside capillary glucose testing in a large teaching hospital: implications for managing point of care testing. Am J Med 1994;97:222–230.

Keffer JH. Economic considerations of point-of-care testing. Am J Clin Pathol 1995;104(Suppl 1):S107–S110.

DeCresce RP, Phillips DL, Hpwanitz PJ. Financial justification of alternate site testing. Arch Pathol Lab Med 1995;119:898–901.

Table 24.
Why Cost Analysis for Bedside Testing Is Complicated

- There are multiple ways to measure cost.
- Most of the cost is for labor.
- Labor costs are hard to measure.
  - Many testing personnel
  - Variable wage rates
  - Personnel performing testing or phlebotomy may change.
  - Testing may be performed in otherwise idle time.
  - Much labor is for training and management.
  - Bedside testing may change productivity.
- Potential benefits are hard to measure.
  - Improved nursing efficiency?
  - Improved patient care?
- Analysis of central lab process may fail to account for all processing steps.
  - Labeling of samples
  - Filling out test requisitions
  - Venous vs. capillary sample collection
  - Retrieval of results
- Costs from decreased accuracy of bedside testing are hard to measure.

2N. Published Cost Analyses

Most published cost analyses report higher costs for bedside testing. Estimated direct costs for bedside glucose testing range from about $4–14 per test while central laboratory glucose testing is estimated to cost about $3–4 per test. These estimates are based on the cost of materials used—about $1—and the time required for testing, quality control, and training multiplied by the wage rate. Most of the reported difference in cost results from a high wage rate for nurses performing the glucose testing versus a low wage rate for phlebotomists collecting samples for central laboratory testing. However, staffing models of many medical care units have changed such that phlebotomy is often the responsibility of nurses or patient care technicians rather than phlebotomists. One study in which the author participated found that it required more nursing time to collect and send a sample to a laboratory than it did to perform the test at the bedside. When this is the case, the labor costs of bedside glucose testing will come out favorable relative to centralized testing.

The author's conclusion in examining published analyses of bedside glucose testing costs is that the reports actually describe cost allocation rather than cost measurement. It is the same process that leads hospitals to conclude that they should charge $50 for a dose of acetaminophen. None of the analyses calculated labor costs in terms of changes in the number of paid employee full-time equivalents (FTE) or paid hours, and these are the numbers that will translate into actual budgetary changes. Adding up the amount of time spent by a worker to perform a task and multiplying by their wage rate does not represent the labor cost because changing from bedside to centralized testing may change the amount of unproductive time and overall productivity. In contrast, to deal with true budgetary impact, address the question of how many new employees you will need to perform specific tasks and what their pay will be or how many staff will be laid off because a new process is more efficient. If your staffing remains constant, you are having no budgetary impact even if you are adding or decreasing the number of tasks that workers must perform. You are only changing productivity.

Table 25.

Published Cost Comparisons of Bedside and Laboratory Glucose Testing

- Most reports claim higher total costs for bedside testing.
- 80-90% of costs is attributed to labor.
- Reports claim higher labor costs for bedside testing.
- Shortcomings of these analyses
    - Studies describe cost allocation for labor, not budgetary impact.
        - They do not describe changes in number of staff or paid hours.
        - They do not measure changes in productivity or unproductive time.
        - Analyses are based on staffing models with blood collection by phlebotomy services. Many hospitals have cut back or eliminated phlebotomy services.

References

Greendyke RM. Cost analysis: bedside glucose testing. Am J Clin Pathol 1992;97:106–107.

Lee-Lewandrowski E, Laposata M, Eschenbach K, et al. Utilization and cost analysis of bedside capillary glucose testing in a large teaching hospital: implications for managing point of care testing. Am J Med 1994;97:222–230.

Winkelman JW, Wybenga DR, Tanasijevic MJ. The fiscal consequences of central vs. distributed testing of glucose. Clin Chem 1994;1628–1630.

20. Comparing Processes of Bedside and Central Laboratory Testing

In comparing bedside and central laboratory testing, it is useful to evaluate all steps in the processes. Sending samples to a central laboratory requires a considerable amount of work for test ordering, sample labeling, generation of a test request or transmittal, packaging of the sample, and transport to the laboratory. Then after the test is done, the result must be retrieved by accessing a computer system, sorting through paper reports, or calling the laboratory. Also, venous blood collection takes substantially longer than capillary blood collection if the patient is a difficult draw or if the patient is on anticoagulation therapy (compression must be maintained on draw sites for up to 10 minutes).

Published cost analyses conclude that less labor is required for central lab testing. This does not appear logical. Breaking the processes down into individual steps (as in Table 26) shows that more than twice as many steps are required for testing by central labs, and every step in the central lab process requires significant labor. The only automated step is the analysis. When nurses are responsible both for phlebotomy and for bedside testing, labor analysis is simplified. A survey the author conducted of 50 nurses at a large hospital found nurses in almost unanimous agreement that it took less of their time to perform the test than to send a sample to the central laboratory and retrieve the results. This subjective response was supported by documenting actual time elapsed when nurses performed phlebotomy and bedside testing.

Bedside testing can be performed in a single visit to a patient's room. This may improve efficiency by allowing the test to be completed in one step rather than in many (collecting samples, waiting one hour, and then going to a nurses' station to sort through lab printouts or access results on a computer terminal). If a laboratory is used, blood draws must be conducted earlier to allow time for results to be returned, and schedules of meals and medications are disrupted if results are not back promptly. If a high volume of glucose tests is sent to a central laboratory, these tests, which tend to fall at the peak time for other testing, can overload the laboratory and delay all test results or require substantial staffing increases. The reduced number of process steps in bedside testing eliminates some errors, such as mislabeling of samples, and improves efficiency, especially when you consider that each step in a process offers an opportunity for delay and error.

Reference

Moultrie CA, Hortin GL, Randolph VR. Comparative costs and labor efficiency of bedside versus central laboratory glucose testing [abstract]. Clin Chem 1997;43:S162.

Table 26.
Comparison of Steps in Bedside and Central Lab Testing

Bedside Testing	Central Lab Testing
Test order noted	Test order noted
Nurse gathers supplies	Test requisition filled out
Nurse walks to room	Label printed
Nurse performs fingerstick	Supplies gathered
Nurse uses glucose meter	Nurse walks to room
Nurse records results	Nurse performs phlebotomy
	Nurse labels sample
	Sample sent to nursing station
	Sample transported to lab
	Sample received by lab
	Sample centrifuged
	Sample analyzed
	Nurse goes to nursing station
	Nurse retrieves result from computer or paper reports
	Nurse records result

Each step in the processes above requires labor. Which process would you expect to require less time?

2P. Estimating Labor Costs vs. Estimating Budgetary Impact of Bedside Glucose Testing

Traditional methods of cost analyses that are applied in published reports do not provide realistic cost estimates that translate into actual budgetary impacts. In order to assess the budgetary impact and more practically estimate the labor cost of a bedside testing program, it is necessary to estimate the impact of a testing program on the number of personnel and amount of overtime. Labor costs will be low if there is available idle time that can be better utilized for performing testing, or if bedside testing actually requires less time than sending samples to a central laboratory. Either of these situations represents opportunities to improve productivity, and increases in labor costs will be less than the actual time spent performing the tasks.

A simplified example of how traditional cost analysis does not provide a useful estimate of the budgetary impact of bedside testing is provided below.

An Example of Traditional Cost Analysis for Labor vs. Estimated Impact on Budgets

A hospital plans to perform 1,000 bedside tests that require about 200 hours to perform.

Two options are evaluated:

1. One FTE will be added to perform the test.
2 Existing staff will perform the tests by using previously unproductive time.

Traditional cost analysis:
Both options have the same labor cost. Each requires the equivalent of about 0.1 FTE to perform the tests. There is no accounting for the 90% of the time that the added employee would spend waiting to perform tests.

Analysis of budgetary impact:
The first option will increase the budget by the wages of one employee.
The second option will not yield any increase in labor expenditure.

The conclusion from this example is that, in many cases, the traditional manner of estimating labor costs does not translate into real changes in budget. The traditional approach is better suited to a factory or reference laboratory setting where work can be organized as a continuous stream, there is little idle time between samples, and work flow has been optimized to achieve highest productivity.

2Q. Estimated Staffing Changes Resulting from Bedside Glucose Testing

It is difficult to come up with general formula to predict the effects of bedside testing on staffing. Healthcare institutions vary greatly with respect to how efficiently their central laboratories process test orders, label tubes, fill out test requisitions, transport specimens, and report results. Additional variables include the degree of laboratory automation, the assignment of personnel to specific tasks, and the usual turn-around time for results. Each laboratory should estimate labor impact by analyzing the average number of samples handled per clerical staff member and technologist in the section performing glucose testing, and the time it takes for glucose tests to be performed. It is important to note whether glucose testing is at a peak time or during a period with excess capacity. Usually, morning fasting glucose levels are needed at the time of peak activity in a hospital laboratory, and the laboratory must be staffed to address the peak workload to avoid delays. Noon or afternoon samples are likely to be drawn at times when hospital laboratories are less busy and can accommodate extra work.

Approximate changes in staffing allowed by bedside glucose testing are listed in Table 27. The table is based on a patient-centered care model where staff on nursing units perform phlebotomy. In this setting, no change in nursing labor is expected to perform beside testing. Phlebotomy and central laboratory glucose testing require approximately the same amount of nursing time as bedside glucose testing. Productivity levels in a laboratory depend significantly on the level of automation. Benchmarking studies and published reports find that each laboratory worker performs approximately 10,000–20,000 tests per year and productivity has been increasing by 5–10% per year. Estimates here for impact of bedside glucose testing on central laboratory staffing therefore are relatively conservative.

References

Benge H, Csako G, Parl FF. A 10-year analysis of "revenues," costs, staffing, and workload in an academic medical center clinical chemistry laboratory. Clin Chem 1993;39:1780–1787.

Hortin GL. Win the productivity battle: the continuing success story of how labs reach new levels of productivity. MLO 1996;28(11):24–30.

Table 27.

Typical Impact of Bedside Testing on Staffing (Nurse-performed phlebotomy)

Nursing staff: No change in FTE

Bedside test management and quality assurance:

- Add 0.5 FTE for up to 50,000 annual patient tests
- Add 1.0 FTE for up to 100,000 annual patient tests
- Add 2.0 FTE for up to 250,000 annual patient tests
- (Number of test sites, meters and testing staff are important variables)

Clerical staff for specimen processing and receiving in laboratory

- Save 1 FTE for every 50,000 annual glucose tests

Technologists for analysis

- Save 1 FTE per every 50,000 annual glucose tests (highly automated lab)
- Save 1 FTE per every 30,000 annual glucose tests (moderate automation)

FTE=full-time equivalent

(Estimates based on references from p. 44 and unpublished benchmarking studies.)

2R. Net Impact of Bedside Glucose Testing on Staffing

Setting up management and quality assurance systems for bedside glucose testing is relatively time consuming, and these functions require about the same amount of time whether test volume is high or low. From the standpoint of labor cost, performing a low volume of bedside glucose testing is relatively undesirable. In clinics or in hospitals with a limited volume of testing on one or two nursing units, bedside glucose testing is likely to have no effect on staffing, although extra time may be required (possible overtime or an extra staff position for a testing coordinator) for test management. The breakeven point for the number of staff required for bedside glucose is about 50 tests per day. At this point, the added staff for managing bedside testing may be offset by reduction of central laboratory staff. For large bedside glucose testing programs in which more than 200 glucose tests per day are performed, there is a potential for net staff and budget reductions.

Many variables affect the impact of bedside glucose testing on staffing:

1. Performance of other bedside tests generally increases test management efficiency if they are combined in the same testing program.
2. Testing management requires more time as the number of test meters, staff, and units performing testing increase (for a given volume of testing).
3. Management becomes less efficient if more than one type of glucose meter is used, and if procedures, training, and policies are not standardized.
4. Staff reductions due to bedside glucose testing decrease if laboratory testing, transport systems, and access to test results are highly efficient.

Table 28.
Estimated Net Staffing Change Due to Bedside Glucose Testing

Annual Test Volume	Approximate Net Change
10,000	Add 0.5 FTE
20,000	No change in FTE
50,000	Save 1.5 FTE
100,000	Save 3.0 FTE
250,000	Save 8.0 FTE

Net staffing changes are calculated based on typical productivity levels in Table 27. Adjust for productivity levels of your own laboratory. The above changes assume that it will be possible to implement staffing changes according to labor demands. In planning, consider whether there is any real opportunity for staff transfers or reductions. Union rules, civil service guidelines, or institutional employment policies may limit the ability to adjust staffing. No true labor savings or increases in labor costs occur unless there are actual changes in staffing through transfer, retirement, or layoff. Reducing labor costs requires long-term planning if disruptions of morale are to be avoided due to layoffs or short staffing.

2S. Materials Costs for Bedside Testing

Equipment and supply costs usually represent a small proportion of total testing costs. Supply costs are usually about $1 per test for a high volume of testing. Test strips for glucose meters usually cost about 50 cents each and some test strips are used for repeat analyses or for control analyses. Other needed supplies include lancets and control materials.

Some strategies to minimize the costs of supplies and meters include:

- standardization of methods and consolidation of purchasing to obtain volume discounts,
- use of purchasing consortia to obtain improved pricing, or
- competitive bidding.

The last option may be desirable for testing programs with extremely high volumes, but it represents more work to prepare a request for bids and to evaluate proposals. Purchasing departments can provide useful information about whether your organization belongs to a buying group and which purchasing options are available, and can assist you with requests for bids. Purchasing supplies in large lots and with the longest possible expiration dates can simplify control programs and reduce the number of test strips needed to check new reagents. Distributors may be able to save a particular lot of supplies and ship from it as needed. Using a meter and reagents that are specific to hospital use may prevent losses due to sending large numbers of test strips home with patients. It may also prevent diversion of meters and supplies for home use by staff members.

If you purchase materials in high volume, it may be practical to have all of the costs for equipment and supplies included as a slight increase in the strip price. This eliminates the need for separate budgeting for capital equipment, which often needs to be planned a year or more in advance, and it provides a means of allocating costs for each analysis.

Table 29.
Controlling Equipment and Supply Cost

Supply costs
- Standardize methods
- Use hospital-specific meters to prevent diversion of supplies
- Buy in large lots

Equipment costs
- Standardize equipment
- Minimize the number of meters in each unit

Purchasing strategies
- Consolidate purchasing to gain volume discounts
- Use purchasing groups to obtain discounts
- Use competitive bidding when volume is very high
- Reagent rental of equipment

2T. The Value of Blood Conservation

A significant benefit of bedside glucose testing vs. centralized testing is reduction in the volume of blood drawn for testing. There is a blood loss of about 0.1 mL (a couple of drops) from a typical fingerstick, whereas blood collection tubes commonly contain from about 2–10 mL of blood. Collection of 5-mL tubes of blood 3 times daily removes 1–2 units of blood per month. This represents a significant loss of iron and other nutrients for a patient with chronic disease. For patients requiring transfusions (at a cost of about $200 per unit), phlebotomy for glucose testing has a blood replacement cost of $5-$10 per day. Also, there is the risk of infectious disease transmission and immunosuppressive effects of transfusion. And patients who do not require transfusions may still need additional nutritional support such as iron supplementation.

The value of blood conservation is even higher for neonates. Transfusions are frequently required to replace the blood neonates lose due to long-term hospitalization. Losses from phlebotomy can represent a severe physiological and nutritional loss for infants with small total blood volumes.

References

Salem M, Chernow B, Burke R, Stacey J, Slogoff M, Sood S. Bedside diagnostic blood testing: its accuracy, rapidity, and utility in blood conservation. JAMA 1991;266:382–389.

Smoller BR, Kruskall MS. Phlebotomy for diagnostic tests in adults: pattern of use and effect on transfusion requirements. N Engl J Med 1986;314:1233–1235.

Table 30.
Value of Blood Conservation

Value of blood that must be replace by transfusion is about 50 cents to $1 direct cost per mL drawn.

- Costs for cross-matching
- Costs for blood products
- Costs for labor and supplies for transfusion
- Increased risk of infectious disease
- Immunosuppressive effects of transfusions

Value is higher for neonates requiring transfusion.

Value of blood not requiring transfusion for replacement is harder to measure.

- Nutritional depletion
- Iron replacement
- Erythropoietin therapy in some cases

2U. Backup Support for Glucose Meters

Often, sites performing bedside glucose testing store one or more extra meters, reagents, and controls as backups. This tends to be wasteful from the standpoint of requiring more meters and more QC tests, tolerating poor maintenance of backups, losing infrequently used meters, and wasting reagents. Adequate backup support via central laboratory testing or other mechanisms can avoid the need to have extra meters at all testing sites.

A trouble-shooting resource such as an on-call service is particularly important for assisting with user and reagent problems. Having extra meters and reagents available from a central location on a 24-hour basis can save on the total number of meters in use. Calibration of backup meters should have been verified within the last 6 months before use. The number of extra meters needed depends on how fast a manufacturer can repair or replace meters that are not functioning properly. Contracts should identify whether there will be any cost for meter replacements, which usually can be sent via express delivery.

In the author's experience, meters experiencing heavy use have required replacement after three to 12 months of use. A testing program should expect to replace or repair some meters, although with ongoing improvements in technology, meter durability will probably improve.

Table 31.
Strategies for Providing Backup Meters

Goals

- Assure availability of testing at all times.
- Minimize number of meters and costs.

Backup mechanisms

- Central laboratory testing as backup
- Trouble-shooting resources on call (especially important for user or reagent problems)
- Backup meters and reagents at a central site
- Backup meters at each site (increases cost)
- Manufacturer's customer support
- Replacement of meters by manufacturer

2V. Strategies to Minimize Costs of Bedside Glucose Testing

The costs of bedside glucose testing depend on the volume of testing. This observation applies primarily to the total volume of testing in a program but also to some degree on the volume of testing at each testing site. The general strategies for minimizing bedside testing cost involve consolidation and standardization. Specifically, consolidate testing into a unified program with the smallest number of high-volume testing sites, and standardize all equipment, training, and procedures.

A key element when evaluating budgetary impact is to use direct measures, such as the increase or decrease in number of paid hours for labor cost, rather than indirect indicators such as charges or time spent performing an activity. To make good management decisions, accurately measure and monitor important aspects of operations such as costs or turn-around times. A summary of cost-saving strategies is provided in Table 32.

Table 32.
Strategies to Minimize Costs

- Evaluate budgetary impact based on changes in paid hours. Use existing idle time and increase productivity.
- Limit glucose meters to units with high testing volumes.
  Costs of training, quality control, and management have a higher cost-per-test on units with a low testing volume.
- Limit the number of meters on each unit.
  Each meter must have daily checks (raises costs).
  Centralize backup support to minimize meters on each unit.
- Standardize—use a single type of meter and test strip.
  Decreased training time, inventory and monitoring costs
  Improved discounts from increased purchase volume
- Centralize purchasing.
  Improved discounts from increased purchase volume
  Use of buying groups to obtain discounts
  Competitive bidding
- Minimize number of lot changes.
  Buy in large volume and seek most distant expiration date.
- Minimize the number of controls run.
  Use meters with QC lockout to assure running of controls
  Minimum requirement is two levels per day
- Use test strips that cannot be used on home glucose meters. (Otherwise many strips may be sent home with patients or taken by staff for home use.)
- Select instruments and train to minimize number of repeats.
- Use vendor support.
  Personnel training
  Assistance with meter evaluation

Clinical Care Issues

3A. Clinical Applications of Blood Glucose Measurement

Blood glucose is one of the most frequently used laboratory tests. In an outpatient setting it is the primary screening and diagnostic test to detect the occurrence of diabetes. There is a growing prevalence of Type II (non-insulin–dependent) diabetes in the U.S. due to obesity and aging of the population. For this reason, glucose is commonly included in test profiles performed in wellness programs or health screening. Glucose is also measured in acutely ill patients and is included in panels that check electrolyte balance. Such profiles are identified by a variety of terms such as "acute care" or "fluid balance profiles" and they include measurement of sodium, potassium, chloride, bicarbonate, and creatinine and/or blood urea nitrogen (BUN), in addition to glucose. The glucose measurement helps provide an overall picture of metabolism, kidney function, and salt balance in patients who are receiving large amounts of intravenous fluids and who are experiencing abnormal kidney function and fluid losses and metabolic changes from stress or disease.

At normal concentrations, glucose comprises a significant component of total osmolality of the serum or plasma fraction of blood—about 5 mOsm/kg out of 300 mOsm/kg. Glucose administered in intravenous solutions is commonly at a concentration of 5% (often termed D5W for 5% dextrose in water), corresponding to 5000 mg/dL in the units commonly reported for blood glucose measurement. High glucose concentrations (in patient samples) can serve as a sensitive indicator for the contamination of blood samples with intravenous fluids: concentrations of glucose as high as 50% (or 50,000 mg/dL) are administered through catheters inserted into large central veins for nutritional support of patients.

Glucose testing for the above applications generally is not performed on glucose meters. For diagnosis of diabetes, most glucose meters are less accurate than is desirable. Also, most glucose meters perform only this single test, and it is more efficient to perform all tests in a panel on the same analyzer. Analyzers that can perform test panels that include various combinations of glucose, electrolyte, and blood gas analyses at the bedside are becoming more available. These analyzers broaden the applications of bedside glucose measurement by providing panels of test results and greater accuracy than is possible with traditional glucose meters. There are tradeoffs: they are found in higher costs for equipment and test cartridges and the need for slightly higher specimen volumes relative to glucose meters.

The major application of glucose meters is to monitor blood glucose concentrations in diabetics. Blood glucose concentrations rise and fall depending on dietary intake, insulin doses, and rates of energy expenditure, and blood may need to be monitored several times daily to determine how to adjust and normalize blood glucose. Patients can use glucose meters to perform self-monitoring at home. Nursing or laboratory staff can measure glucose concentrations with either a glucose meter or with laboratory analyzers. Glucose meters also may be used where there is a desire to rapidly check whether a patient has low blood sugar. This may include diabetics being treated with insulin, newborns who often experience hypoglycemia until after they begin feeding, patients with severe liver failure, or patients with other rare metabolic disorders such as insulin-producing tumors or inborn errors of metabolism.

Table 33.
Clinical Applications of Blood Glucose Testing

Laboratory glucose testing

- Screening and diagnosis of diabetes
- Monitoring fluid and electrolyte balance of acutely ill patients
- Monitoring glucose levels in aptients with metabolic disorders

Glucose meter or central laboratory testing

- Monitoring of diabetes therapy
- Detection of hypoglycemia
    - Insulin-dependent diabetics
    - Newborns
    - Hypoglycemic disorders
    - Liver failure
    - Insulinoma
    - Inborn errors of metabolism

3B. Blood Specimen Collection

Specimen collection is a key part of accurate blood glucose monitoring. Even the most accurate instruments will not provide valid results on an inappropriately collected specimen. Usually, blood glucose monitoring is associated with capillary collections drawn by finger sticks, but a number of other means of specimen collection may be applied such as heel punctures for babies, venipuncture, drawing through peripheral or central venous lines, or through arterial lines. If venous specimens are to be used, the meter must not be affected by a low oxygen concentration. Many meters rely on reactions of the enzyme glucose oxidase, which uses oxygen to react with glucose. Some of these meters may be affected by low oxygen content of venus blood. Capillary specimens have a composition similar to arterial blood, with a high oxygen concentration and slightly higher glucose concentration than venous specimens.

Usually, specimens are applied immediately to glucose meters and no anticoagulant is used. If specimens are collected into anticoagulants such as heparin in blood gas syringes or in green-topped tubes or fluoride/oxalate in gray-topped tubes, the meter needs to be checked for compatibility with the anticoagulant. Individual meters may differ in their acceptance of anticoagulated specimens.

Samples drawn through intravenous (i.v.) lines must be flushed with a preliminary blood draw before specimen collection. In order to prevent dilution of blood by solutions infusing through the i.v. line, a volume equal to at least twice the volume within the line must be drawn. The volume within most i.v. lines is 2 mL (cc) or less. Observe the volume drawn before blood reaches the syringe and double it. This applies to solutions such as normal saline or half-normal saline that do not contain glucose. Glucose (dextrose) solutions given through lines are commonly administered as D5W, which is 5% glucose in water. This is equivalent to 5000 mg/dL, and even slight contamination with i.v. fluids containing glucose will cause erroneously high blood glucose values. If a glucose solution is infusing, avoid drawing blood specimens from the line. If there is no other blood access available, a volume of blood equal to six times the line volume should be drawn before a specimen is collected. For babies, protocols are commonly employed to minimize blood loss by reinfusing blood removed to flush lines. This requires close attention to aseptic technique and avoidance of clotting and introduction of air.

Table 34.
Blood Specimen Collection for Glucose Meters

- Capillary specimens— similar to arterial blood
  High oxygen concentration, slightly higher glucose than venous
    Fingersticks
    Heel punctures
- Venous draws—low oxygen concentration
  Check on compatibility of meter
  Avoid anticoagulants unless compatibility is checked.
    Venipuncture
    Intravenous lines
      Need to withdraw at least two volumes of blood before specimen collection to avoid dilution with i.v. fluids
      High risk of erroneous glucose values if glucose is infusing
- Arterial draws—high oxygen concentration and slightly higher glucose than venous specimens
  Blood gas specimens—check on compatibility of meter with heparin
  Arterial lines
    Need to flush similar to i.v. lines
    High risk of erroneous glucose values if glucose is infusing

References

Baer DM. Drawing from i.v. lines. MLO 1998;30(2):12.

Clapham MCC, Willis N, Mapleson WW. Minimum volume of discard for valid blood sampling from indwelling arterial cannulae. Br J Anaesth 1987;59:232-235.

3C. Capillary Blood Collection

Take universal precautions when collecting blood. Thus, wear a fresh pair of gloves during specimen collection. Perform fingersticks along the side of the fingertip. The routine procedure is to clean the fingertip with an alcohol wipe, allow the site to air dry, perform a puncture with a sterile lancet, blot the first small drop with a cotton ball, then allow a drop to form and apply it to a test strip without direct contact between the finger and test strip. The site of collection must have good circulation. (Values may be falsely low if the patient has poor circulation to the site due to hypotension, shock, vascular disease, or cold hands.) Blood flow to the finger may be stimulated by warming the hand before specimen collection. Avoid excessively squeezing the finger to force out a drop of blood, because squeezing forces out an increased amount of tissue fluid, which has a lower glucose concentration.

In a healthcare setting, it is desirable to use spring-loaded safety lancets rather than lancets with an open needle. The needle or blade that performs the puncture is retracted into a protective cover except for the instant when skin puncture is performed. The spring-loaded lancets are more expensive, but they reduce the risk of accidental needlestick injury to staff and patients, and the spring-loaded lancets help avoid either excessively deep or shallow skin puncture. All parts that come into contact with the patient should be disposable. Reusable platforms and spring-loaded devices where only a needle is changed have been described as a source of transmission of viral hepatitis between patients. Laser devices are available for skin puncture, but the lasers are larger and more expensive than lancets and appear to cause similar amounts of pain.

Heel puncture of babies should occur around the edge of the heel, not in the middle where bone is closest to the surface. Pediatric lancets should be used to provide the proper depth of puncture. Excessive depth of puncture leads to a risk of contacting bone and introduction of osteomyelitis, an infection, in the heel bone.

Table 35.
Capillary Blood Collection

Procedure

- Use universal precautions—wear gloves
- Check for adequate circulation
- Warm hand if necessary
- Clean site with alcohol wipe or soap and water
- Allow to dry
- Perform skin puncture
- Blot off first small drop
- Allow drop to form without excessive squeezing
- Touch drop but not finger to test strip
- Dispose of blood-contaminated lancets and waste appropriately

Recommendations

- Use single-use spring loaded lancets
  - Protection from accidental needlesticks
  - Consistency
- Avoid lancing devices with reusable parts
  - Infection control issues
- High cost laser devices not yet of proven benefit

References

Burge MR, Costello DJ, Peacock SJ, Friedman NM. Use of a laser skin perforator for determination of capillary blood glucose yields reliable results and high patient acceptability. Diabetes Care 1998;21:871–873.

Gerberding JL. Management of occupational exposure to blood-borne viruses. N Engl J Med 1995;332:444–451.

Hortin GL, Utz C. Lancets for capillary blood collection. Am Clin Lab 1996;15(4):8–9.

Howanitz PJ, Schifman RB. Phlebotomists' safety practices: a college of american pathologists Q-Probes study of 683 institutions. Arch Pathol Lab Med 1994;118:957–962.

3D. Diabetes Mellitus

Diabetes is a disorder in the body's use of the sugar glucose. Tissues are unable to take up glucose into their cells appropriately due to a deficiency of the hormone insulin or to a relative lack of response to the hormone. As a consequence, the concentration of sugar circulating in blood increases. Sustained high levels of glucose lead to tissue injury and a diverse range of complications including impaired kidney function; eye damage that can leading to blindness; damage to blood vessels with resulting poor circulation to extremities and increased risk of heart attack and stroke; and damage to nerves, causing loss of peripheral sensation and dysfunction of the autonomic nervous system. Treatment is aimed at lowering blood glucose levels so as to delay or prevent the onset of diabetic complications as well as acute metabolic disorder.

There are four major categories of diabetes (see Table 36). Type I, or insulin-dependent diabetes, usually has onset in childhood and is characterized by a severe deficiency of insulin due to destruction of the beta cells in the pancreas that synthesize insulin. Type II, or non-insulin–dependent diabetes, occurs in middle age or later and is associated with obesity.

Table 36.
Types of Diabetes

- Type I (Insulin-dependent, juvenile)
  Lack of insulin due to beta cell destruction
- Type II (Non-insulin–dependent, adult-onset)
  Associated with obesity, insulin resistance
- Gestational
  Occurring during pregnancy
- Secondary
  Related to medications such as corticosteroids or to pancreatic destruction or endocrine diseases such as acromegaly and Cushing's Disease

References

Expert Committee on the Diagnosis and Classification of Diabetes Mellitus. Report of the expert committee on the diagnosis and classification of diabetes mellitus. Diabetes Care 1997;20:1183–1197.

The Diabetes Control and Complications Trial Research Group. The effect of intensive treatment of diabetes on the development and progression of long-term complications in insulin-dependent diabetes mellitus. New Engl J Med 1993;329:977–986.

3E. Diagnosis of Diabetes

Revised criteria for the diagnosis of diabetes were developed in 1997. The new criteria lowered the threshold value for diagnosis of diabetes from a fasting plasma glucose level of 140 mg/dL (1985 guidelines from WHO) to 125 mg/dL. Note that diagnosis is based on the concentration in the plasma fraction of blood (the liquid components after all cells are removed by centrifugation). An estimated 12% of the U.S. population over the age of 40 has diabetes. The number of diabetics is expected to grow with the aging of the population and increased incidence of obesity.

Most glucose meters are not well suited to diagnosing diabetes. The accuracy of measurements is usually about ± 20%, which means an uncertainty of about ± 25 mg/dL at the cutoff for diagnosis on a fasting sample. In addition, many glucose meters may give a value for whole blood glucose that is about 10% lower than the glucose concentration in the plasma fraction of blood.

Reference

Expert Committee on the Diagnosis and Classification of Diabetes Mellitus. Report of the expert committee on the diagnosis and classification of diabetes mellitus. Diabetes Care 1997;20:1183–1197.

Table 37.
New Criteria for Diagnosis of Diabetes

- Plasma glucose > 125 mg/dL for a fasting subject (no caloric intake for at least 8 hours)
- Plasma glucose > 200 mg/dL on a random sample for a person with symptoms of polyuria, polydipsia, and weight loss
- Plasma glucose > 200 mg/dL two hours after drinking 75 grams of glucose in water for a glucose tolerance test

3F. Insulin Treatment of Diabetics

Administration of insulin doses needs to be timed very carefully with respect to meals. The timing of insulin doses depends on the type of insulin used, blood glucose levels, and sites of injection. Usual guidelines for insulin doses before meals state that doses of rapid-acting insulin should be given within 15 minutes of a meal and doses of short-acting insulin should be given about 30 minutes before meals. The need for tight coordination of glucose measurements, insulin doses, and meals is one of the major reasons for use of glucose meters. Delays in test results will delay insulin doses and meals. Peak action of insulin doses occurs at an interval about 2-4 times the length of time required for onset of action.

Table 38.
Time for Onset of Insulin Action

Type of Insulin	Onset of Action
Rapid-acting (Lispro)	15 minutes
Short-acting (Regular)	30 minutes
Intermediate-acting (NPH)	180 minutes
Long-acting (Lente)	360 minutes

References

American Diabetes Association. Insulin administration. Diabetes Care 1998;21(Suppl 1):S72–S75.

Davidson MB. Diabetes mellitus: diagnosis and treatment. Philadelphia: WB Saunders, 1998.

Skyler JS. Insulin therapy in type I diabetes mellitus. In: DeFronzo RA, ed. Current therapy of diabetes mellitus. St. Louis, MO: Mosby, 1997.

3G. Hypoglycemia (Low Blood Sugar): A Common Emergency in Insulin-Dependent Diabetics

Although the goal of intensive therapy of insulin-dependent diabetes is to keep blood glucose levels from going too high, this is a careful balancing act. It is also harmful if blood glucose levels become too low. Hypoglycemia is an acute emergency because starvation of the brain for glucose can cause seizures, loss of consciousness, and, if uncorrected, death. Even with careful management, as in the Diabetes Control and Complications Trial, more than half of patients undergoing intensive treatment have severe episodes of hypoglycemia. The trial reported that six severe (requiring assistance of another individual) episodes occurred per year of patient treatment and about 25% of the episodes involved loss of consciousness or seizures. In a large study reviewing the cause of death in insulin-dependent diabetics, hypoglycemia was identified as the cause of death in 4% of subjects. Occurrence of hypoglycemia serves as the overall limiting factor in how aggressively blood glucose levels can be controlled in diabetics. Mild hypoglycemia is very common during intensive insulin treatment; it may be present (yet unapparent to the treated individual) during as much as 10% of a typical day. Fear of hypoglycemia sometimes serves as a psychological barrier to tight control of blood glucose.

Reference

The Diabetes Control and Complications Trial Research Group. The effect of intensive treatment of diabetes on the development and progression of long-term complications in insulin-dependent diabetes mellitus. New Engl J Med 1993;329:977–986.

Cryer PE. Hypoglycemia: pathophysiology, diagnosis, and treatment. New York: Oxford University Press, 1997.

Cox DJ, Irvine A, Gonder-Frederick L, Nowacek G, Butterfield J. Fear of hypoglycemia: quantification, validation, and utilization. Diabetes Care 1987;10:617–621.

Table 39.
Significance of Hypoglycemia for Care of Insulin-Dependent Diabetics

- A common side effect of insulin therapy
- Often not obvious to diabetic
- Limiting factor in efforts to reduce blood glucose
- Potential for coma, brain damage, and death
- Avoidance requires frequent testing of blood glucose and careful adjustment of insulin doses and food intake

References

Cryer PE. Hypoglycemia: the limiting factor in the management of IDDM. Diabetes 1994;43:1378–1379.

Cryer PE. Hypoglycemia: pathophysiology, diagnosis, and treatment. New York: Oxford University Press, 1997.

Nordfeldt S, Ludvigsson J. Severe hypoglycemia in children with IDDM. Diabetes Care 1997;20:497–450.

3H. Cause of Hypoglycemia

In normal individuals, severe hypoglycemia is rare due to minute-to-minute control of the secretion of insulin, the hormone that has the major role in lowering blood glucose by stimulating uptake of glucose by tissues. Insulin secretion for the pancreas is shut off when blood glucose levels fall to about 80 mg/dL. This helps prevent glucose levels from falling lower.

However, insulin-dependent diabetics do not have this physiological control of insulin levels. Their insulin levels depend on the injected dose and the rate of absorption from the injection site. Insulin levels can remain high even as blood glucose levels fall below normal, and this promotes further decreases in blood glucose. As a result, appropriate insulin dosing requires frequent measurement of blood glucose levels and adjustment for varying food intake or metabolic demands. Bedside glucose monitoring may assist in maintaining a regular schedule of glucose monitoring and avoiding delays in meals (waiting for test results) that could disrupt the balance between food intake and insulin dosing.

The body uses several mechanisms for preventing blood glucose from falling too low, including release of hormones such as adrenaline and cortisol which raise blood sugar concentrations. However, these safeguards can be overwhelmed if insulin levels are too high and there is limited sugar uptake from food. Hypoglycemia can occur when the insulin dose is too high relative to food intake. Potential causes of hypoglycemic episodes are listed in Table 40.

Avoidance of hypoglycemia requires careful regulation of insulin and food intake and adjustment for any change in metabolic needs due to acute illness or changes in exercise patterns. Some medications such as beta blockers (commonly used to treat high blood pressure) may increase the risk of hypoglycemia by inhibiting responses to adrenaline. However, increased risk is mainly a factor in older nonselective beta blockers; so-called cardiac-specific beta blockers do not appear to substantially increase that risk. Also, several oral drugs that diabetics take to lower glucose levels are much less frequently associated with severe hypoglycemia than is insulin treatment.

References

Shorr R, Ray WA, Daugherty JR, Griffin MR. Anithypertensives and the risk of serious hypoglycemis in older persons using insulin or sulfonylureas. JAMA 1997;278:40–43.

Edwards TH, Braunstein GD, Davidson MB. Glyburide-induced hypoglycemia in an elderly patient: similarity of first-generation and second-generation sufonylurea agents. Mt Sinai J Med 1985;52-644–647.

Trovati M, Rurzacca S, Mularoni E, et al. Occurrence of low blood glucose concentrations during the afternoon in type II (non-insulin–dependent) diabetic patients on oral hypoglycemic agents: importance of blood glucose monitoring. Diabetologia 1991;34:662–667.

Table 40.
Causes of Hypoglycemia in Diabetics

- Insulin does too high
- Rapid absorption of insulin dose (rate depends on site and exercise)
- Taking insulin at wrong time
- Changes in food intake
  - Meals delayed
  - Meals skipped
  - Vomiting
- Changes in glucose use by the body
  - Acute illness
  - Changes in exercise
- Doses of oral diabetes medication too high
- Liver failure
- Other medications
  - Nonselective beta blockers

31. Risk Factors for Hypoglycemia in Insulin-Dependent Diabetes (Type I)

Several factors can identify diabetics who are at highest risk for developing hypoglycemia. Obviously, changes in insulin dosage or acute illness with major changes in metabolism and food intake due to anorexia or vomiting are important short-term factors. Some less obvious long-term factors associated increased risk of hypoglycemia are intensive insulin treatment, complete loss of endogenous insulin production, and duration of insulin-dependent diabetes of more than 6 years. Intensive insulin treatment, with its goal of preventing high blood glucose levels, leaves little margin for error in regulating blood glucose levels, and hypoglycemia is a common consequence. Some Type I diabetics continue to make small amounts on insulin. Even small amounts of insulin production under physiological control, helps to avoid excessive insulin levels, the body's release of insulin is shut down below glucose levels of 80 mg/dL. Finally, insulin-dependent diabetics who have had their disease for >6 years are often unable to sense when they are becoming hypoglycemic, and their body's ability to release hormones that act to restore blood glucose levels diminishes.

Table 41.
Risk Factors for Hypoglycemic Episodes

- Intensive insulin treatment
- Complete lack of endogenous insulin production
- Diabetes for more than 6 years

References

Cryer PE. Hypoglycemia: the limiting factor in the management of IDDM. Diabetes 1994;43:1378–1379.

Cryer PE. Hypoglycemia: pathophysiology, diagnosis, and treatment. New York: Oxford University Press, 1997.

3J. Hypoglycemia in Non-Insulin–Dependent Diabetics (Type II)

In patients with Type II (non-insulin–dependent) diabetes, several factors have been considered to contribute to the risk of hypoglycemia, including treatment with oral diabetes medication and advanced age. Ther are reports of hypoglycemia from oral medications, however, a recent study suggests that the risk of hypoglycemia is low in elderly patients treated with the sulfonylurea class of oral diabetes medication. No episode of hypoglycemia was noted in 156 episodes of prolonged fasting by elderly patients treated with sulfonylureas. Increases in blood adrenaline levels appear to help maintain blood glucose in Type II diabetic patients. Most oral agents have a slower onset of action than insulin and have a normal duration of action of about 12–24 hours.

References

Burge MR, Schmitz-Firoentino K, Fischette C, Qualls CR, Schade DS. A prospective trial of risk factors for sulfonylurea-induced hypoglycemia in type II diabetes mellitus. JAMA 1998;279:137–143.

Gaster B, Hirsch IB. The effect of improved glycemic control on complications in type II diabetes. Arch Intern Med 1998;158:134–140.

Edwards TH, Braunstein GD, Davidson MB. Glyburide-induced hypoglycemia in an elderly patient: similarity of first-generation and second-generation sufonylurea agents. Mt Sinai J Med 1985;52-644–647.

Trovati M, Rurzacca S, Mularoni E, et al. Occurrence of low blood glucose concentrations during the afternoon in type II (non-insulin–dependent) diabetic patients on oral hypoglycemic agents: importance of blood glucose monitoring. Diabetologia 1991;34:662–667.

Table 42.

Lower Risk of Hypoglycemia in Non-Insulin–Dependent Diabetics

- Medications are less potent at lowering blood glucose.
- Counter-regulatory responses such as release of adrenaline are more effective than in Type I diabetics.

3K. Symptoms of Hypoglycemia (Low Blood Sugar)

It is important for staff who work with diabetics to recognize symptoms of low blood sugar in order to treat it early and thereby avoid severe episodes of hypoglycemia. Bedside glucose monitoring may have a role in rapidly confirming suspected hypoglycemia and in diagnosing hypoglycemia in patients with longstanding diabetes, who often lose normal physiological responses to hypoglycemia.

When blood glucose levels fall below normal, the body activates several pathways to restore glucose levels. One response is to activate the sympathetic nervous system, which stimulates the adrenal gland to release adrenaline (also known as epinephrine) into the circulation. Adrenaline makes the heart beat harder and faster and causes increased nervousness and a sense of anxiety. Another branch of the sympathetic system stimulates sweating and tingling of the extremities by releasing acetylcholine from nerve endings. These symptoms may be recognized by patients or their caregivers as early signals of hypoglycemia. Unfortunately, nerve damage is one long-term complication of diabetes, and normal responses and ability of diabetics to sense low blood sugar may be lost over time. The hormones cortisol, glucagon, and growth hormone also are released in response to low blood glucose and help to maintain blood sugar levels, but acute responses of these hormones are not associated with symptoms that can be recognized.

Other symptoms of hypoglycemia relate to the direct effect of low blood sugar on brain function. The brain depends on glucose as a primary energy source. Consequently, low blood sugar starves the brain of a food source and leads progressively to changes in behavior, confusion, loss of consciousness, and death. Sometimes behavior can become unusual or even bizarre during hypoglycemic episodes and moods may be variable. Confusion or changes in mental status should be recognized as potential signs of hypoglycemia in a diabetic. Rapid treatment is important for avoiding progressively severe consequences.

References

American Diabetes Association. Insulin administration. Diabetes Care 1998;21(Suppl 1):S72–S75.

Davidson MB. Diabetes mellitus: diagnosis and treatment. Philadelphia: WB Saunders, 1998.

Macheca MKK. Diabetic hypoglycemia: how to keep the threat at bay. Am J Nurs 1993(April):26–30.

Bosboom WMJ, Frijns CJM, Van Gijn J. Yelling attacks and wasted hands. Lancet 1996;348:238.

Dizon AM, Danese R, Hoogwerf GJ. Two patients with neuroglycopenia. Cleveland Clin J Med 1998;65:82–86.

Table 43.
Symptoms of Hypoglycemia

Sympathetic nervous stimulation

- Increased adrenaline release
- Shakiness/trembling
- Rapid pulse
- Pounding heart (beating stronger)
- Anxiety or nervousness
- Release of acetylcholine
- Sweating
- Hunger
- Tingling of extremities (paresthesias)

Direct effects of glucose starvation on the brain

- Confusion
- Unusual or silly behavior
- Mood disturbance
- Sensation of warmth
- Weakness and fatigue
- Seizures
- Coma

3L. Blood Glucose Level that Indicate Hypoglycemia

Plasma glucose levels <50 mg/dL are often considered to indicate hypoglycemia. However, the exact glucose levels at which a particular individual will experience symptoms varies. Slowing of mental function may occur below 50 mg/dL and levels falling below 40 mg/dL may lead to loss of consciousness. Many laboratories treat glucose values of below 40 mg/dL as critical values that require immediate reporting.

Normal individuals have a plasma blood glucose level of 70-105 mg/dL when no food has been eaten for several hours (fasting glucose level). For a couple of hours after eating or drinking sugar-rich beverages, plasma glucose rises modestly as high as 150 mg/dL. Current targets for glucose control in diabetics are a plasma glucose level of 80-120 mg/dL before meals and a level of 100-140 mg/dL at bedtime.

Target levels provide a relatively small margin for error before symptoms of mild hypoglycemia occur. In some individuals, symptoms of hypoglycemia may appear at glucose levels of 50-60 mg/dL, although certain symptoms, such as sweating and a sense of nervousness, may be blunted or occur at lower threshold levels in diabetics. Lack of awareness of hypoglycemia in diabetics can be a significant problem, because it prevents early treatment by food or sugar intake before mental function is affected. Usual threshold concentrations for different responses to falling plasma glucose concentrations are listed in Table 44. These threshold values do not appear to be constant, however, even within the same individual. The body may reset threshold levels at different points if an individual frequently has low or high blood glucose levels.

Table 44.
Typical Threshold Values for Responses to Glucose Concentration

Glucose Concentration	Symptoms or Physiological Response
70-105 mg/dL	Normal fasting range
< 80 mg/dL	Decreased insulin secretion
60-70 mg/dL	Increased adrenaline, glucagon, cortisol, and growth hormone
40-50 mg/dL	Slowing of mental function
< 40 mg/dL	Confusion Loss of consciousness Seizures

Note that thresholds may vary somewhat between individuals and may be reset by recurrent high or low glucose values.

References

Cryer PE. Hypoglycemia: the limiting factor in the management of IDDM. Diabetes 1994;43:1378–1379.

Cryer PE. Hypoglycemia: pathophysiology, diagnosis, and treatment. New York: Oxford University Press, 1997.

Emancipator K. Critical values: ASCP practice parameter. Am J Clin Pathol 1997;108:247–253.

3M. Treatment of Hypoglycemia

Hypoglycemia usually can be treated very easily if it is recognized early. It can be reversed within minutes by ingestion of a source of sugar such as fruit juice, soft drinks, sugar tablets, crackers, candy, or other food: have the patient eat foods or drink liquids or sugar tablets containing 20 grams of glucose. Unconscious patients may require intravenous glucose or an injection of glucagon, which rapidly raises blood sugar levels by stimulating the breakdown of starch in the body. Recovery of glucose levels and relief of symptoms should occur within 15 minutes.

Usually, treatment of hypoglycemia produces complete recovery, although permanent brain injury has been described in rare cases. Animal studies emphasize the need for rapid correction of hypoglycemia. When plasma glucose in monkeys was reduced to 20 mg/dL, permanent injury to the nervous system occurred within several hours.

References

Macheca MKK. Diabetic hypoglycemia: how to keep the threat at bay. Am J Nurs 1993(April):26–30.

Kahn KJ, Myers RE. Insulin-induced hypoglycemia in the non-human primate. I. Clinical consequences. In Brierly JB, Meldrum BS, eds. Brain hypoxia, clinics in developmental medicine. London: Heineman, 1971;185–193.

Gold AE, Marshall SM. Cortical blindness and cerebral infarction associated with severe hypoglycemia. Diabetes Care 1996;19:1001-1003.

Table 45.
What To Do about a Hypoglycemic Episode

Give sugar orally if subject is conscious

Sugar, juice, or soft drinks

Diabetics at risk should carry sugar tablets

Give IV glucose or IM glucagon if subject is unconscious

50 mL (cc) of 50% glucose intravenously

Or a shot of 1 mg glucagon IM or subQ

Check blood glucose level to confirm hypoglycemia

Monitor symptoms

Improvement within 15 minutes if due to hypoglycemia

3N. Loss of Consciousness by a Diabetic

Hypoglycemia needs to be considered as the first possibility when an insulin-dependent diabetic rapidly loses consciousness. Treatment is simple and can prevent permanent injury to the nervous system. For this reason, diabetics taking insulin often wear Medic Alert bracelets or necklaces or other indications of their condition in case they require assistance.

With inadequate insulin treatment, diabetics may go into diabetic ketoacidosis or hyperosmolar coma, which are characterized by very high rather than low blood glucose levels. These disorders usually have gradual onset. Rapidly checking blood glucose levels can distinguish whether the patient has extremes of low or high blood glucose. If no abnormality of blood glucose levels is noted, and the patient does not respond to treatment within 15 minutes, other causes of loss of consciousness need to be considered such as fainting, stroke, drug intoxication, traumatic injury, or shock.

Table 46.
Coma in Insulin-Dependent Diabetics

Low blood sugar
- Insulin dose too high
- Skipped or delayed meals

High blood sugar
- Diabetic ketoacidosis
- Hyperosmolar coma

Unrelated to blood sugar level
- Fainting
- Alcohol or drug intoxication
- Stroke
- Shock
- Traumatic injury

Technical Issues

4A. How Glucose Meters Work

Glucose meters use several technologies to measure glucose on whole blood specimens. Non-wipe systems that use a photometer for light detection sieve out the red blood cells, and the fluid elements of blood move to a bottom layer of the strip containing the reagents that react with glucose. Specificity for glucose is provided by the enzymes glucose oxidase, glucose dehydrogenase, or hexokinase. Enzyme reactions are coupled to the oxidation of a dye that changes color. Color changes are measured by reflectance photometry from the bottom of the test strip.

Recently, electrode technology has been applied to glucose measurement. This eliminates the need for a photometer and for separating the intensely colored red cells from the plasma. Enzyme reactions of glucose oxidase or glucose dehydrogenase are coupled to reactions that release electrons and the glucose meter measures the current produced.

Reference

Chmielewski SA. Advances and strategies for glucose monitoring. Am J Clin Pathol 1995;104(Suppl 1):S59–S71.

Table 47.
Methods for Glucose Measurement

Glucose oxidase photometric methods

$$\text{Glucose} + O_2 \xrightarrow{\text{Glucose Oxidase}} \text{Gluconic acid} + H_2O_2$$

$$H_2O_2 + \text{Reduced Dye} \xrightarrow{\text{Peroxidase}} \text{Oxidized dye} + H_2O$$

Color change from oxidation of the dye is measured.

Glucose dehydrogenase (GDH) photometric methods

$$\text{Glucose} + NAD^+ \xrightarrow{\text{GDH}} \text{Gluconolactone} + \text{NADH}$$

Absorbance of NADH can be measured or the enzyme can be coupled to reduction of another dye.

Hexokinase photometric methods

$$\text{Glucose} + \text{ATP} \xrightarrow{\text{Hexokinase}} \text{Glucose-6-phosphate} + \text{ADP}$$

$$\text{G-6-P} + NAD^+ \longrightarrow \text{6-Phospho-gluconolactone} + \text{NADH}$$

$$\text{NADH} + \text{Oxidized dye} \longrightarrow \text{NAD} + \text{Reduced dye}$$

The color change from dye reduction is measured.

Glucose dehydrogenase (GDH) electrode method

$$\text{Glucose} + Fe^{3+}(CN)_6 \xrightarrow{\text{GDH}} \text{Gluconolactone} + Fe^{2+}(CN)_6$$

$$FE^{2+}(CN)_6 \longrightarrow Fe^{3+}(CN)_6 + e^-$$

The current generated is measured. Similar electrode reactions can be coupled to reactions of glucose oxidase

4B. Advantages of New Methods

There have been several reasons for applying new technologies to glucose meters. Use of enzymes other than glucose oxidase has been explored in order to decrease the effect of oxygen concentration on glucose measurements. Low oxygen concentrations can affect measurements with glucose oxidase because oxygen, together with glucose, is a substrate of the enzyme. Using other enzymes improves the suitability of a glucose meter for measuring venous specimens with low oxygen concentrations and decreases effects of hypoxia and altitude. These features are desirable in glucose meters for hospital use where it may be useful to test samples drawn from veins.

The use of an electrode rather than a photometric method decreases the need to clean optical windows in the analyzer. This decreases the maintenance of meters and eliminates a source of potential error—accumulation of dirt or blood in the light path. Also, there is decreased potential for interference from colored substances such as hemoglobin, bilirubin, or high concentrations of lipids.

Applications of new technologies to meters have sought also to reduce specimen volume requirements and to provide error codes when inadequate specimen is tested.

Table 48.
Reasons for New Meter Technologies

- Less effect of blood oxygen concentration on results
  - Allows analysis of venous blood
  - Use on hypoxic patients
  - Use at high altitude
- Elimination of photometers
  - Decreased maintenance (no cleaning of light path)
  - Decreased chance for optical interference
    - Dirt or blood in light path
    - Colored substances in plasma such as bilirubin, free hemoglobin, lipemia
- Decreased specimen volume
  - Easier specimen collection
  - Less chance for errors from short samples
- Error detection for inadequate specimens
  - Avoid reporting erroneous results

4C. Common Mistakes with Glucose Meters

Despite improvements in glucose meters, there are a variety of ways to make errors. Some of them relate to improper sample collection. Others relate to improper operation of meters or to the presence of interfering materials in blood.

In troubleshooting problems with bedside measurement of blood glucose, evaluate these potential sources. When there is a sizeable disagreement between values from a glucose meter and a laboratory instrument, the glucose meter is usually considered to be in error. However, potential sources of error in central laboratory testing and true differences between whole blood and plasma glucose concentration or between capillary and venous sites should be recognized (see sections 4E, Differences between Whole Blood and Plasma Glucose Methods, and 4F, Differences between Fingerstick and Venous Samples). Some sources of error in central laboratory glucose measurement are sample or patient misidentification, data entry errors, or changes in glucose during specimen transport.

New glucose meters have decreased the opportunities for operator error. Previously, variation in technique for wiping strips was a major source of error, and application of too small of a drop of blood would result in falsely low values. Now, most meters have some means of detecting when insufficient specimen was applied, although detection of this error is not perfect. For some meters, touching fingers to the blood application point can affect results. Test strips contain enzyme reagents that are susceptible to breakdown if strips are exposed to moisture or to increased temperatures. Most test strips are packaged in vials that contain a desiccant to prevent moisture buildup. Leaving vials open can make their performance deteriorate. If this problem is likely to occur in a particular hospital setting, obtain some test strips in single-use packages.

The keys to preventing errors and to identifying them when they do occur are thorough training of personnel, effective quality assurance programs, and knowledge of potential sources of error or true difference between methods.

Table 49.
Errors in Glucose Measurements

Specimen collection errors

- Draw from an IV line with glucose solution in it (5% dextrose is 5000 mg/dL)
- Contamination of sample with alcohol (wiping a finger and not allowing it to dry)
- Drawing from a site with poor circulation
- Excessive squeezing of a finger (dilution of blood with tissue fluid)

Operator errors

- Improper storage of test strips (leaving uncapped)
- Inadequate sample volume
- Use of wrong lot number for calibration
- Touching surface of test strip with fingers
- Improper maintenance of equipment
- Dropping and physical damage to meters

Interferences

- Extremely high or low hematocrits
- Reducing or oxidizing substances
- Low oxygen concentration in blood (some meters)

4D. Why Glucose Meter Results Do Not Exactly Match Central Laboratory Results

Results from glucose meters often do not exactly match glucose results from a central laboratory for samples drawn from the same patient at the same time. In a large national survey of bedside glucose test performance in small hospitals, ~10–15% of test results from a glucose monitor differed by more than 20% from results for the same sample when it was analyzed in a central laboratory. Part of this may relate to the lower accuracy and precision of meters. Calibration and regulation of reaction temperatures and measurement of samples for analysis are less precise. Also, manual procedures have variability in operator technique. In many studies, most difference between bedside and central laboratory results are considered errors in the bedside test method. However, this is not always true; some difference in results is expected based on differences in sample type (capillary versus venous samples) and analysis of whole blood at the bedside vs. analysis of centrifuged blood. Delays between bedside and central laboratory analysis may also allow glucose levels to change between measurements, and interferences may affect bedside and central laboratory methods differently.

Reference

Novis DA, Jones BA. Inter-institutional comparison of bedside blood glucose monitoring program characteristics, accuracy performance, and quality control documentation. Arch Pathol Lab Med 1998;122:495–502.

Table 50.
Reasons for Differences between Bedside and Central Laboratory Glucose Results

- Higher variability (imprecision) of bedside methods
- Variability in operator technique
- Different sample types
    Capillary samples versus venous
- Whole blood versus plasma
- Delays between bedside and central lab analysis
- Different interferences for bedside and central lab methods

4E. Differences between Whole Blood and Plasma Glucose Methods

Whole blood has a glucose concentration about 10% lower than that of plasma or serum fractions. Glucose equilibrates between the plasma and the intracellular space of red blood cells, but about one-third of the space within the red blood cells is taken up by hemoglobin. The exact difference between whole blood and plasma depends on the hemoglobin content of blood. As the hemoglobin concentration increases, whole blood concentration will be progressively lower than the plasma concentration. Comparison studies of methods measuring whole blood concentrations vs. methods of measuring plasma glucose should correct for hemoglobin concentration or hematocrit. Another consequence of hematocrit effects on whole blood glucose is that care needs to be taken to thoroughly mix whole blood samples; otherwise settling of red cells can lead to unpredictably low or high hematocrits in the portion of a sample tested.

At very high hematocrits, typically over 55%, another factor that may affect the performance of some glucose meters is the high viscosity of samples. High viscosity may interfere with the penetration of blood into a multilayer test strip, and the decreased amount of plasma water may affect the ratio of reagents to water. Most glucose meters identify an upper limit for hematocrit that has been verified to deliver acceptable performance. Patients who often have markedly elevated hematocrits include newborns, polycythemic, and severely dehydrated subjects.

Plasma is not considered an acceptable specimen for most glucose meters because the low hematocrit will yield different results than whole blood for the reasons stated above. Similarly, proficiency-testing specimens ideally should resemble whole blood rather than plasma or simple aqueous solutions of glucose. Proficiency-testing programs may have a choice of several specimen types, and it is important to choose a specimen type that is compatible with your meter.

References

Maser RE, Butler MA, DeCherney GS. Use of arterial blood with bedside glucose reflectance meters in an intensive care unit: are they accurate? Crit Care Med 1994;22:595–599.

Ramamurthy RS, Berlanga M. Postnatal alteration in hematocrit and viscosity in normal and polycythemic infants. J Pediatr 1987;110:929–934.

Somer T, Meiselman HJ. Disorders of blood viscosity. Ann Med 1993;25:31–39.

Young DS, Bermes EW. Specimen collection and processing; sources of biological variation. In Burtis CA, Ashwood ER, eds. Tietz textbook of clinical chemistry, 2nd ed. Philadelphia: WB Saunders, 1994; 58–101.

Table 51.
Difference between Whole Blood and Plasma Glucose

- Whole blood has a lower glucose concentration
    - Average difference is 10%
    - Difference depends on hematocrit (due to volume displacement by hemoglobin)
- Unpredictable effects of extremely high hematocrits
    - Increased sample viscosity
    - Decreased plasma water for diluting reagents
- Check manufacturer's recommended hematocrit range
- Potential problem specimens
    - Plasma
    - Samples with high hematocrits
        - Newborns
        - Polycythemic patients
        - Severely dehydrated patients

4F. Differences in Glucose Concentration between Fingerstick and Venous Samples

Blood collected from fingersticks or heel punctures can have a different glucose concentration than blood collected through a needle inserted into a vein, the routine method for collection of laboratory blood specimens. Fingersticks yield blood from the small blood vessels (capillaries), and this blood has a composition similar to arterial blood as long as there is good circulation of blood to the site. Arterial blood has higher blood glucose and oxygen concentrations than does venous blood: in venous blood, oxygen and glucose are extracted by tissues in the capillary beds connecting the arterial and venous circulation. Thus, the following discussion on differences between capillary and venous blood applies equally to arterial vs. venous specimens.

In a fasting state with low insulin concentrations, the difference between capillary and venous specimens drawn at the same time from the same subject is usually about 2–10 mg/dL (higher in capillary specimens). However, following a meal, when glucose and insulin levels rise, there is a much larger difference between capillary and venous blood because insulin stimulates tissue uptake of glucose. As an example, when samples are drawn 30 minutes after drinking a sugar solution for a glucose tolerance test, an average difference of 45 mg/dL is found between capillary and venous specimens. Since reference ranges and medical interpretation of blood glucose values are based on the plasma concentration in venous specimens, fingerstick collections are most suitable for fasting specimens. Also, if comparisons are to be made between glucose meter and central laboratory glucose results, both tests should be done on a capillary specimen or, if glucose meter results from capillary specimens are to be compared with results on venous specimens, the specimens should be drawn while the patient is in a fasting state. Otherwise, it is hard to know whether differences are due to differences in measurement or in specimen type.

Another factor that should be considered for fingerstick specimens is that the site of collection needs to have good blood circulation. If there is poor circulation to the site due to peripheral vascular disease, shock, or other causes, blood glucose may be depleted due to stagnant flow. Fingerstick collections for laboratory testing are not appropriate for a patient in shock or for one who has other causes of poor circulation.

References

Burgi W. Oraler glucosetoleranztest: unterschiedlicher verlauf der kapillaren und venosen blastungskurven. Schweiz Med Wschr 1974;104:1698–1699.

Kupke IR, Kather B, Zeugner S. On the composition of capillary and venous blood serum. Clin Chim Acta 1981;112:177–185.

Larsson-Conn U. Differences between capillary and venous blood glucose during oral glucose tolerance tests. Scan J Clin Lab Invest 1976;36:805–808.

Table 52.
Differences in Glucose Concentration between Capillary and Venous Blood

- Fasting state
  - Insulin levels and tissue glucose uptake are low.
  - Capillary blood has a glucose concentration 2-10 mg/dL (higher than venous blood).
- After meals
  - Insulin levels and tissue glucose uptake are high
  - Capillary blood has a higher glucose concentration than venous blood—averaging as much as 45 mg/dL higher.
- Consequences
  - Fingerstick specimens are best suited for fasting measurements.
  - Method comparisons should analyze the same specimen or compare fasting specimens.

4G. Effects of Oxygen Concentration on Glucose Measurement

The higher oxygen concentration of arterial and capillary blood relative to venous blood is evident from the bright red color of oxygenated hemoglobin in the capillary blood and the much darker red of deoxyhemoglobin in venous samples. Oxygen concentration has an important effect on the performance of some, but not all, meters that use the enzyme glucose oxidase to measure glucose. The enzyme reactions in these meters rely on oxygen for the oxidation step. Some glucose meters do not perform well with venous specimens or other specimens containing low oxygen concentrations. (Information to this effect is usually indicated by the manufacturer and states that the meter is not approved or suitable for use with venous specimens.) Glucose meters that do not rely on glucose oxidase for glucose measurements are not affected by oxygen concentration. Laboratory glucose methods also are not affected, although this is a potential interference with whole blood analyzers (used at the point-of-care or in laboratories) that rely on electrodes containing glucose oxidase as an active element.

Other situations in which oxygen tension of capillary specimens may be decreased include operation at high altitude, collection of specimens from sites that have poor circulation, or in patients with hypoxia. Altitude effects usually require elevations above 5000 feet. Sites for fingerstick collections should have good blood circulation, otherwise not only is the blood oxygen level likely to be low, but also other nutrients such as glucose are likely to be depleted. Results from fingerstick glucose measurements may be misleadingly low in patients who are in shock or who are hypotensive because circulation of blood to the extremities is likely to be inadequate.

References

Halloran SP. Influence of blood oxygen tension on dipstick glucose determinations. Clin Chem 1989;35:1268–1269.

Kurahashi K, Maruta H, Usuda Y, Ohtsuka M. Influence of blood sample oxygen tension on blood glucose concentration measured using an enzyme-electrode method. Crit Care Med 1997;25:231–235.

Zaloga GP. Beware of errors in glucose measurement. Crit Care Med 1997;25:212.

Table 53.
Effects of Oxygen Concentration on Glucose Measurement

- Capillary and arterial blood have high oxygen content and venous blood has low oxygen content.
- Oxygen is a reactant for meters using the enzyme glucose oxidase to measure glucose.
- Low oxygen concentration interferes with glucose measurement in some meters. Such meters should not be used
  - For venous samples
  - For samples from hypoxic patients
  - At high altitudes
  - For specimens from poorly perfused sites
- Meter selection should consider the need for venous testing and evaluate the meter's appropriateness for that use.

4H. Potential Interferences with Glucose Measurements

Glucose meters are more prone to interferences than are routine central laboratory testing methods. This is because glucose meters perform a measurement directly on an undiluted sample of whole blood and potential interferents are present at a high concentration. Most glucose analyses in a laboratory are performed after centrifugation of a sample and dilution of the specimen in reagents for measurements; dilution of potential interferents usually lessens their effect. The separation of the plasma or serum fractions of whole blood also allows visual detection of high concentrations of colored substances such as bilirubin or free hemoglobin (from lysis of red cells in the body or during collection) and turbidity from high concentrations of lipids. This identifies specimens that will potentially interfere with photometric measurements of glucose.

At high concentrations, compounds that act as reducing agents—such as vitamin C, uric acid, cysteine, glutathione, L-DOPA, bilirubin, or dopamine—can interfere with the oxidation of dyes used for color reactions in photometric glucose meters. These compounds can affect the number of electrons released during glucose analysis in glucose meters based on electrode technologies. Central laboratory methods for measuring glucose are not completely immune to interference from these reducing compounds, but the magnitude of interference is usually smaller.

Sugars other than glucose are another potential interference. Methods using the enzyme glucose oxidase are usually quite specific for glucose. Methods using other enzymes such as glucose dehydrogenase may react with other sugars such as galactose, which is produced by the breakdown of the lactose abundant in milk. This is rarely a problem because galactose levels usually are much lower than those of glucose; however, patients with a rare inborn error of metabolism called galactosemia may be an exception.

Other potential interferences from low oxygen concentration and high hematocrit have been detailed previously (see section 4E, "Differences between Whole Blood and Plasma Glucose Methods). Very low or high plasma protein concentrations in specimens may also affect results. For this reason, simple water solutions of glucose may not read as expected, and proficiency-testing materials will not perform equivalently on all glucose meters.

Table 54.
Potential Interferences with Glucose Meters

Compounds with potential interference

- Reducing substances
  - Vitamin C
  - Uric acid
  - Dopamine and L-DOPA
  - Bilirubin
  - Cysteine
  - Glutathione
- Sugars other than glucose
  - Galactose

Low oxygen concentration

High hematocrit

Abnormal protein concentrations

4I. Replacing Old Meters

There have been rapid technological developments in glucose meters. Each testing program staff must evaluate whether new technologies are of value for their setting. Potential benefits must be weighed against investment of time and effort that will be devoted to evaluating a new system and retraining its users.

Some potential benefits of changing glucose meters are to improve the quality of results (for example, switching from a meter that requires wiping of test strips to a no-wipe meter); to improve data systems and test management; to allow testing of additional specimen types such as venous or neonatal specimens; to take advantage of lower pricing; to consolidate all testing into use of a single type of meter; or to obtain improved customer support. A history of problems with test quality, poor user acceptance, high costs, or poor support from a vendor might also lead to consideration of a new meter.

In most cases, the primary goals are to improve test quality, to improve data collection, and to reduce cost for equipment, reagents, and training through standardization of methods within a hospital or healthcare system.

Table 55.
Why Replace Old Meters

- Improve test quality
  - Decreased operator dependence
  - Improved precision
  - Decreased interferences
- Accept additional sample types
  - Venous, arterial, neonatal
- Increase analytical ranges
- Improve data systems
  - Better result and QC tracking
  - Lockouts
- Lower costs
- Improve system durability

4J. Selecting a Glucose Meter

Meter technology has been changing so rapidly that it is useful to contact vendors to obtain information about their most recent product offerings. A number of factors to consider are listed in the accompanying table.

An important point to consider is that you are not just buying a meter; you are also buying ongoing support. It is hard to measure the level of support that you will receive based on brochures and promises. The best measure of expected support for training, troubleshooting, shipping of supplies, and repair of meters is to talk to other local customers about how well the company has delivered. Sales representatives should be able to provide a list of references, and networking with other sites that perform bedside glucose testing can provide valuable information.

Meter durability is another issue that is difficult to measure in a brief evaluation, and the experience of customers who have used a meter for a year or more provides a better measure. A potential drawback with newly released products is that some problems with maintenance or durability may not show up until the products have been on the market for six months to a year.

In most healthcare organizations there are advantages to selecting a meter that is designed for hospital rather than home use. In general, meters for hospital use accept a wider range of specimen types (such as venous as well as capillary specimens), have more sophisticated data systems, and prevent potential losses by being "incompatible" with home-use systems (thereby decreasing the likelihood that staff or patients will take them home). Hospital-use meters do not use the same test strips as home meters. If a low volume of testing is to be performed and testing will be performed under a certificate of waiver, low meter cost and whether a meter has a "waived" classification will be important factors. If a high volume of testing is performed, the cost of test supplies usually exceeds that of meters; these costs may actually be included in the reagent price as a reagent rental.

Table 56.
Choosing a Test System

Major considerations

- Customer support
  - Check local references
- Suitability for sample types
  - Samples from newborns?
  - Testing of venous samples?
- Performance characteristics
  - Accuracy and precision
  - Interferences
  - Analytical range
- Ease of use
- Tolerance of operator variability
- Data system
- Costs
- Infection control issues
- Required maintenance
- Durability
- Hospital use or home use meters

4K. Evaluating a Glucose Meter

Evaluation needs to include the usual analytical parameters such as precision, linear range, and accuracy (comparison vs. a reference method). However, one way in which the evaluation for a bedside instrument should differ is that there should also be field testing by personnel who will be performing the tests. A common practice is to have the most experienced and skilled personnel perform evaluations, but it is important when using bedside instruments to see whether personnel with the lowest level of training and experience will have problems using the meter. The challenge is to determine whether the least expert user and most disorganized nursing unit will provide accurate results.

Potential users should try the system and have input in the evaluation process. A simple questionnaire may be used to collect feedback from a number of users. Feedback from testing personnel will help identify practical problems and indicate how hard the system is for them to learn. Nursing personnel are more likely to identify problems such as infection control issues if meters are prone to blood contamination and difficult to clean. Another benefit of seeking input from testing personnel relates to the issue of change: there will be greater acceptance of a new system if testing personnel know that they have had input into decision-making.

If your care setting includes atypical patient populations such as babies, dialysis patients, transplant patients or others with high or low hematocrits or potential interferences, check whether the meter will work for your patients. For correlation studies, try to test over a wide range of glucose values and patient types. If possible, perform tests in duplicate to identify the rate of random errors. Ideally, the reference method should be performed on the same specimen as the glucose meter. If the reference method cannot be performed on a capillary specimen, correlation studies should be done on fasting patients to minimize the differences between capillary and venous glucose concentrations.

As data systems become more complicated, they need to be evaluated more thoroughly to check whether features such as security codes and quality control lockouts work as intended. Report generation should be tested to see that desired reports can be produced.

Reference

Nichols JH, Howard C, Loman K, Miller C, Nyberg D, Chan DW. Laboratory and bedside evaluation of portable glucose meters. Am J Clin Pathol 1995;103:244–251.

Voss EM, Bina DM, McNeill LD, Cembrowski GS. Determining acceptability of blood glucose meters: design of meter evaluations. Lab Med 1996;27:534–537.

Voss EM, Bina DM, McNeill LD, Johnson ML, Cembrowski GS. Determining acceptability of blood glucose meters: evaluating a blood glucose testing system. Lab Med 1996;27:679–682.

Table 57.

Evaluating a Glucose Meter

- Analytical performance
  - Precision
  - Linear range
  - Accuracy—correlation vs. reference method
    - Check over a wide range of glucose values and patient types.
  - Effect of interferences
- User evaluation—opportunities for input
  - How difficult to learn
  - Ease of use
  - Suitability for sites of intended use
  - Infection control issues
- Operation of data systems
  - Data storage
  - Security codes
  - Lockouts
  - Data uploads from meters
  - Report generation

4L. Data Systems for Glucose Meters

Assessment of the data systems of glucose meters has become an increasingly important element in meter selection. When manual log sheets are employed, it is difficult to all ensure that all results are recorded. Pages are lost or some staff forget to record results when they are pressed for time. A data system allows the utilization of glucose testing to be tracked for each nursing unit and each patient, and it serves to document each test result, time of testing, test operator, reagent lot number, and comment codes as well as maintenance logs and quality control results. Each field, such as patient and operator ID, must have the necessary number of digits.

Meters should be able to store more than a month's accumulated testing, if you are planning to do monthly summaries. There should be a simple means of uploading results of individual meters via a network or into a laptop computer that can be taken to the nursing units. The central data system should be able to generate summary management and quality control reports for each meter, nursing unit, and for the overall program

Lockout features can be set to require the adequate performance of quality control testing before patient testing can be performed, and access codes can prevent testing by unauthorized users. Testing performance and quality control analyses by each user can be tracked. Notes of frequent repeats or problems can identify operators who need additional training. Many meters can accommodate bar code readers to read patient armbands, staff identification cards, or other information; this speeds data entry and decreases error rates.

Data systems have been one of the most rapidly evolving elements of glucose meters. Information on current systems capabilities should be obtained from vendors, because information in this or any other publication rapidly becomes out of date. The major challenge for data systems is to develop a simple and inexpensive means of networking meters so that results can be transmitted rapidly to central laboratory information systems. This requires simple data ports, an established network for communication, an interface with laboratory information systems, and an editor to check patient identifications and locations (to avoid sending results to the wrong location). Such networking capability for glucose meters is currently available. The other option for entering results into laboratory information systems has been manual data entry, which becomes labor intensive if testing volume is high.

References

Hortin GL, Utz C, Gibson C. Management of information from bedside testing. MLO 1995;27(1):28–33.

Ingram-Main R, Wright J, Kiechle FL. Data management programs for bedside glucose testing. Lab Med 1994;25L784–789.

Nichols JH. Data management's role in point-of-care testing. Advance/Lab 1997:6(7):8–12.

Jacobs E, Laudin AG. The satellite laboratory and point-of-care testing: integration of information. Am J Clin Pathol 1995;104(Suppl 1):S33–S39.

Table 58.
Potential Benefits of Data Systems

- Elimination of manual logs
- Complete records and documentation of all patient testing
- Avoidance of transcription errors in data transfer
- Tracking of all users
- Lockout of unauthorized users
- Tracking of quality control results
- Lockout of meters when quality control fails
- Improved access to patient records
- Improved tracking of utilization
- Workload recording
- Improved ability to bill or account for costs
- Decrease time to generate summary reports
- Consolidation of all records
- Merging of results with other laboratory records

Quality Assurance

5A. Quality Assurance of Bedside Glucose Testing

Quality assurance programs include a number of elements, some of which are required by federal standards, state regulations, or standards of inspecting agencies such as the College of American Pathologists (CAP) and the Joint Commission on Accreditation of Healthcare Organizations (JCAHO). There is good evidence that the quality of testing can be improved by implementation of quality assurance programs. This is true, as described by Stahl and others, particularly in the case of the physician's office laboratory. In addition to complying with all elements of the quality assurance program, it is important to maintain adequate records of performance and to organize them so that they can be retrieved for inspections or reviews of operations.

During laboratory inspections, there is particular interest in documentation of corrective actions taken in response to problems such as out-of-range quality control values or proficiency test results that are out of acceptable limits. Appropriate actions are the best indication that problems are recognized and remedied. Elements of a quality assurance program for bedside glucose testing differ little from other testing; the major difference is simply application of the program to a larger number of testing sites and personnel.

All policies and procedures for testing and quality assurance activities should be in writing and should be reviewed and signed by the designated laboratory director. Some accrediting agencies, such as CAP, require these to be in a specific format and to include specific information. Policies should describe all elements of the testing program and should include elements listed in Table 59.

Reference

Stahl M, Brandslund I, Iversen S, Filtenborg JA. Quality assessment of blood glucose testing in general practitioners' offices improves quality. Clin Chem 1997;43:1926–1931.

Walker EA. Quality assurance for blood glucose monitoring: the balance of feasibility and standards. Nursing Clin N Am 1993;28:61–70.

Jurf JB, Outlaw EG. Designing and implementing a comprehensive quality assurance program for bedside glucose testing. Diabetes Educator 1993;19:105–108.

National Committee for Clinical Laboratory Standards. Ancillary (bedside) blood glucose testing in acute and chronic care facilities: approved guideline. Villanova, PA: NCCLS, 1994.

Table 59.
Elements of Quality Assurance Programs

- Written policies and procedures
- Personnel standards
- Training programs
- Competency testing
- Quality control testing
- Proficiency testing
- Critical values policies
- Reporting of results
- Equipment maintenance
- Equipment evaluation
- Tracking of incident reports
- Quality monitors
- Documentation of corrective actions
- Retention of all records at least 2 years

5B. Quality Control Testing

CLIA regulations dictate that for all quantitative tests, at least two levels of control materials—materials with set concentrations of an analyte such as glucose—should be checked daily. Two or three levels of these control materials, with high, low, or normal glucose concentration, are checked. The concentration of glucose and expected variation in measurement of the control material is determined by performing multiple measurements with normally operating meters. Control limits (acceptable ranges of values) are set so that severe deviations from normal meter measurements are recognized when they occur. For glucose meters, the control limits for a single control analysis is usually set at the mean ±3 standard deviations. If the control measurement is outside this range, the control material is retested or other corrective action is taken until control values are within acceptable limits.

Manufacturers often list assigned control ranges on the vials of control materials. These ranges should be confirmed early during the use of new control material to check that the mean and range are appropriate. Incorrect ranges can frustrate testing personnel, because out-of-range control measurements will increase, requiring repeat testing and investigation. Control ranges should be checked when new lots of test strips are put into service to see whether the control range requires adjustment. It is desirable to obtain reagents and controls with the longest possible stability and to avoid frequent lot changes, both because these entail additional work and because frequent lot changes make it difficult to identify trends in results.

Table 60.
Quality Control Testing

- A minimum of two levels of control daily for each meter
- Common patterns of control testing for glucose meters:
  - Two levels—low and high
  - Three levels—low, normal, high
- Control limits: mean ±3 standard deviations
  - Manufacturers' range measured at factory
  - Verify for your meters
    - Check ranges for new lots of control material
    - Check ranges when test strip lot changes
- Try to avoid frequent changes in lots of reagents and controls.

References

Ehrmeyer SS, Laessig RH. Regulatory requirements (CLIA '88, JCAHO, CAP) for decentralized testing. Am J Clin Pathol 1995;104(Suppl 1):S40–S49.

Jones BA, Howanitz PJ. Bedside glucose monitoring quality control practices: a college of american pathologists Q-Probes study of program quality control documentation, program characteristics, and accuracy performance in 544 institutions. Arch Pathol Lab Med 1996;120:339–345.

Westgard JO, Bawa N, Ross JW, Lawson NS. Laboratory precision performance: state of the art versus operating specifications that assure the analytical quality required by Clinical Laboratory Improvement Amendments proficiency testing. Arch Pathol Lab Med 1996;120:621–625.

5C. Review of Quality Control Results

No patient testing should be performed when the meter's control results are out of range: the control test is indicating that the meter is probably not producing accurate results. When reviewing control results, check that this policy is adhered to strictly and review the type of corrective action taken when control results failed. Operators with frequent failures of quality control testing probably need retraining. Ideally, control testing should be rotated among various personnel, because, for bedside testing, it is important to control for operator variability as well as for instrument and reagent variability. The control limits for glucose meters of ±3 standard deviations are relatively liberal compared with multi-rule control limits used in central laboratories, and the broader tolerance in variation of results from glucose meters probably contributes to lower accuracy and precision of bedside glucose meter results. Tightening control limits for meters would improve accuracy slightly but at the cost of more repeats of control analyses and additional troubleshooting.

Most of the time when a control result is out of range it does not indicate a meter problem. More often it indicates a problem with the control solution or an operator error. For example, the wrong calibration may have been used for the lot of strips tested. Operators need to be educated about possible sources of error in control testing so that they can recognize the problem and repeat the analysis to obtain acceptable results. Common causes of out-of-range control values are listed in Table 61.

Printing monthly summary statistics of the mean and coefficient of variation for each meter provides a quick measure of whether a meter is performing with typical accuracy and precision. Review of Levey-Jennings plots of all control values allows more sensitive detection of trends in control values, and it may identify meters that consistently read one or two standard deviations from the expected means. These meters, although not consistently failing the 3 standard deviations criteria, are not providing the most accurate values and should be serviced or replaced. Directors and coordinators of testing should provide documentation of at least monthly review of control results.

Reference

Westgard JO, Bawa N, Ross JW, Lawson NS. Laboratory precision performance: state of the art versus operating specifications that assure the analytical quality required by Clinical Laboratory Improvement Amendments proficiency testing. Arch Pathol Lab Med 1996;120:621–625.

Table 61.
Causes of Out-of-Range Control Values

- Use of wrong control material (switching high and low controls)
- Deterioration of controls
  - Cross-contamination of different levels of control
  - Evaporation
  - Microbial growth
  - Dirt or other contaminants
- Operator error
  - Inadequate drop size
  - Application of bubbles
  - Touching strip surface with bottle or finger
  - Use of wrong calibration codes
  - Use of wrong lot of reagents
  - Other instrument-specific errors
- Deterioration of test strips
- Instrument problems
  - Optical path or electrodes need cleaning
  - Physical damage to instrument
  - Instrument malfunction

5D. Limitations of Control Testing of Glucose Meters

Laboratory workers consider control testing the central and most important element of laboratory quality assurance programs. Unfortunately, it does not succeed quite as well in the bedside testing environment as it does in the central laboratory. First, glucose meters are unit-use devices. Each test strip is a unique test device and control testing of one does not guarantee that the next strip will perform equivalently: the next strip could have been damaged during shipping or storage or loading in the meter. Second, when operator technique may affect results for a manual method such as use of glucose meters, controls performed by one operator may not assure proper testing by another operator. Third, performing control analyses is not the same as performing patient testing, and control testing does not check for correct application of blood to the test strip. Fourth, control testing at a central location does not guarantee that the meter will perform exactly the same at another location with different temperature, humidity, lighting, and other environmental differences. Finally, control testing is relatively more expensive for devices such as glucose meters where a large number of analyzers are in use and each one must undergo control testing. The ratio of control results to patient results is higher.

These limitations suggest that there are other elements of quality assurance programs that assume greater importance for bedside testing. Personnel training and checking of competence assume greater importance when controls cannot identify problems with testing. Equipment selection for ease of use, durability, and simple maintenance is important. The ability of analyzers to detect malfunctions such as obstructions of light paths or low battery power and to detect inadequate specimens are important assurances of test quality.

A controversial point concerns to what degree electronic checks of meter function can replace measurement of control materials. Electronic checks have been applied to a number of test devices, and they lower the time and cost associated with control programs. Appropriate applications and regulatory acceptance of this alternative approach have not been resolved fully.

Reference

Hortin GL. Beyond traditional quality control: how to check costs and quality of point-of-care testing. MLO 1997;29(9):31–37.

Table 62.
Limitations of Control Testing for Glucose Meters

Limitations

- Testing of one test strip does not assure that others will work (each is a unit-use device).
- Control testing by one staff member does not ensure equivalent performance by other staff members.
- Technique for application of controls is different than for application of patient specimens.
- Control testing at one location does not control for performance under different environmental conditions.
- Control testing is expensive because many meters must be checked daily.

Consequences

- Other quality assurance measures are of increased importance for bedside glucose testing.
- There is interest in alternative means of checking meter operation.

5E. Personnel Training

Personnel training is one of the most critical factors in ensuring testing quality, and it is often one of the most difficult. It is challenging to schedule and train large numbers of staff, and to convey the necessary (and consistent) information when staff motivation, training, and background may be highly variable. Staff need to know not only the fundamentals of operating meters and performing tests, but also policies for safety, specimen collection, quality control testing, critical values, and necessary documentation. Procedure manuals should be available to testing personnel and they should become familiar with the information that the manuals contain. It is important to maintain records of the training and competency checks of personnel.

Vendors will often assist with staff training. This training would need to be supplemented with any policies that are unique to your testing program. Training videos that supplement hands-on experience with a meter, followed by a thorough policy review specific to your institution, can provide a basic introduction to using bedside glucose testing. Staff competency should then be tested.

Table 63.
Personnel Training

- Review of policies
  - Contents of procedure manual
  - Safety
  - Specimen collection
  - Quality control testing
  - Critical values
  - Documentation
- Test procedure
  - Use of meter
  - Control testing
  - Maintenance
  - Basic trouble-shooting
  - Whom to call for help
- Checking competency

5F. Competency Checks

During initial training, staff should be tested to see that they really have learned the policies and procedures that they will need to apply for use of glucose meters. Document the competency check of each staff member, and ensure that competency is checked periodically, at least annually, to verify that staff members have not forgotten information that they need to know. JCAHO and CAP require competency testing but have not prescribed the exact manner in which it should be performed. Various options can be used: administration of written tests, analysis of test or control specimens, or observations of staff members while they perform tests. One advantage of written tests is that they can be used directly for documentation. A checklist of needed skills and knowledge should be prepared, and strategies for reviewing all critical elements should be developed. Supervisors or coordinators of testing may need to review their competency over a broader spectrum of skills and knowledge than staff who perform routine testing.

Reference

Berte LM. Training verification for lab personnel: a new guide. MLO 1995;27(1):38–42.

Nordenson NJ. Strategic competency assessment: a plan to minimize extra work. MLO 1997(October);29:50–53.

Table 64.
Methods of Performing Competency Checks

- Written tests
  - Policies
  - Basic knowledge
  - Testing procedures
- Performance of test
  - Knowledge of operation
  - Develop manual skill
  - Reinforcement of learning
  - Control testing to provide a measure of appropriate technique
- Direct observation of testing
  - Sample collection technique
  - Testing technique
  - Performance of controls

5G. Calibration Verification

For glucose meters that are classified under the Clinical Laboratory Improvement Act (CLIA) as moderate complexity (which includes many for hospital use), regulations specify that there should be a linearity check of each meter before it is put into use and that calibration should be verified at least twice a year. However, the CLIA Interpretative Guidelines have not been clear about this requirement and indicate that testing programs should follow manufacturers' guidelines.

The point of this requirement is to check that each meter works over its entire glucose-measuring range. Meter checks should include at least high-, mid-, and low-level test materials of known concentration. Sets of calibration verifiers that contain five levels of glucose are available from manufacturers; plotting programs or graph paper that help to describe whether results are within allowable limits are also available. When meters undergo substantial service or repair, their calibration should be checked before they are put into service to ensure that they are operating correctly. All records of meter checks should be kept for at least 2 years.

CLIA and JCAHO have no specific requirements for meters classified as "waived testing" other than those described in the manufacturers' directions for use. CAP standards require calibration verification for all quantitative methods and do not recognize a different standard for waived testing.

References

Baer DM. Linearity of glucose meters. MLO 1997;29(9):20.

HCFA. Survey procedures and interpretative guidelines for laboratories and laboratory services. January 1993, page C-107.

US Code of Federal Regulations. 42 CFR Part 493.1213(b)(2)(1) and Part 493.1202(c)(3).

5H. Proficiency-Testing Programs

Proficiency testing serves as a valuable means of checking the accuracy of test results. Sets of specimens are distributed to laboratories across the country and results are collected to determine how your performance compares to that of other laboratories using the same or different meters. There is evidence that proficiency testing, quality assurance, and laboratory regulations actually can improve the quality of testing; one such example is a physician's office laboratory (POL). POLs that have had less extensive quality assurance programs have performed more poorly on proficiency testing. After enrollment in proficiency testing, these laboratories progressively improve their proficiency test results over time.

There are a number of CLIA-approved proficiency-testing programs. Some proficiency-testing providers are listed in the following table. Contact these programs for specifics about cost and availability of specimens for testing of glucose meters.

One of the distinctions that CLIA makes for waived testing, which includes testing with many glucose meters, is that proficiency testing is not required. However, proficiency testing is a valuable tool for verifying testing accuracy through comparison with other laboratories performing the same tests. Also, many accrediting agencies such as CAP and JCAHO require proficiency testing, and CLIA requires proficiency testing for certified laboratories performing non-waived tests.

One loophole with respect to glucose meters is that, if a laboratory has multiple methods for measuring glucose, it is able to designate a primary method for which proficiency testing results are graded for acceptability. If proficiency testing then is performed on meters, results are reported but not graded. When results for glucose meters are reported as a primary method, only results for two meters are graded. This does not mean necessarily that proficiency samples should be checked on only two meters, however. Proficiency testing of all meters provides a means of comparing the performance of all instruments on the same specimens.

Table 65.
Partial Listing of CLIA-Approved Proficiency-Testing Providers

Providers	Phone Numbers
Accutest, Inc.	800-356-6788
American Academy of Family Physicians (AAFP)	800-274-7911
American Association of Bioanalysts (AAB)	847-981-7662
College of American Pathologists (CAP)	800-323-4040
State departments of health—Idaho, Maryland, New Jersey, New York, Ohio and Pennsylvania	————
Wisconsin State Laboratory of Hygiene	800-462-5261

References

Crawley R, Belsey R, Brock D, Baer DM. Regulation of physicians' office laboratories: the Idaho experience. JAMA 1986;255:374–382.

Hurst J, Nickel K, Hilborne LH. Are physicians' office laboratory results of comparative quality to those produced in other laboratory settings? JAMA 1998;279:468–471.

Pontius CA. Managing proficiency testing analysis. Vantage Point 1998:2(5):10–11.

Stull TM Hearn TL, Hancock JS, Handsfield JH, Collins CL. Variation in proficiency testing performance by testing site. JAMA 1998;279:463–467.

5I. Performing Proficiency Testing

Proficiency-testing programs must provide at least three shipments of specimens per year, which have from two to five specimens for testing in each shipment. Laboratories must have satisfactory performance on 80% or more of test results in order to be considered acceptable. Failure to achieve 80% requires investigation and exception reports describing corrective actions.

Participating laboratories must perform testing in the same way that they do for patient samples. This means that the same personnel who normally perform testing, whether they are nurses, clerks, or medical technologists, perform the proficiency testing. (If samples require reconstitution and mixing before analysis, this can be performed by special technical staff, since it is not a routine part of testing. Also, special designated staff may distribute specimens and tabulate results on reporting forms.) Testing should not be performed multiple times to obtain an average or "best" result for reporting or be referred to another laboratory in order to obtain a better result. Violations in these CLIA-defined rules for proficiency testing may result in loss of certification or other penalties.

It is important to pick the right vendor and specimen type for proficiency testing of glucose meters. Various compounds are added as stabilizers to the blood specimens used for proficiency testing, and these differences in the sample composition or matrix vs. normal whole blood have variable effects on different glucose meters. Scoring of results is based on comparisons within peer groups, that is, laboratories that use the same type of meter. Acceptable results are within 3 standard deviations of the group mean. Your meters should be of the same type as those used by other hospitals participating in the proficiency program or you will have no peer group for comparison—your performance will be compared with other meters that may provide different results. Check with vendors of glucose meters about recommendations for the most suitable proficiency surveys for their meter, and check with the proficiency program that there is a large peer group for comparison.

Table 66.
Picking a Proficiency Provider

- Seek recommendations about vendors.
  - Meters react differently to the abnormal sample matrix.
  - Which program is used most often for a meter?
- Call testing program.
  - Find out how many sites with the same type of meter are enrolled.
  - A critical point is that evaluation is vs. a peer group.
    - The proficiency program needs many participants with the same meter.

5J. Critical Values

Current standards of laboratory practice and many accrediting agencies such as the CAP and JCAHO require specific policies for reporting laboratory test results that are highly abnormal and that represent potentially life-threatening conditions. Testing for blood glucose is one example of tests in which very low or very high values may be considered critical. Note that a variety of other terms have been used when referring to critical values: panic values, life-threatening values, or alert values. A specific hospital or laboratory may be accustomed to using a different term, but the meaning is the same.

Policies for critical care have several purposes. First, they aim to ensure that appropriate patient care is delivered when there might be a need for acute treatment. Second, they address standards of care and laboratory regulations. Finally, they are a component of quality assurance programs: extreme laboratory values sometimes indicate instrument, reagent, or operator failures; sample contamination (e.g., with i.v. fluids); or interferences from other causes.

There are four main elements to policies for critical values. First, put the policy in writing—it must be written down, reviewed, distributed, and included in training. Second, identify which laboratory measurements and threshold values are critical for the specific care setting. Third, describe what to do if a critical value occurs. This includes specific measures to verify a result, such as by repeating an analysis; procedures for communicating the test result; and personnel who need to be notified of the test result. Usually, results must be reported directly to clinical staff—nurses or physicians. If a central laboratory performs the test, this might involve contacting the clinical staff by telephone. When nurses perform a test at the bedside, information may be given directly to the user (unless policy requires notification of a physician). If bedside testing is performed by phlebotomists, laboratory staff, or nursing assistants, policies should identify whom they should notify about critical values. Finally have a mechanism to document occurrence and communication of critical values. For bedside glucose testing, data systems for many glucose meters allow entry of comment codes that can acknowledge recognition of critical values. For central laboratory glucose results, laboratory reports include notations of critical values and often indicate who was contacted with the results. Laboratories maintain logs of calls for critical values either in data systems or in handwritten log books.

References

Emancipator K. Critical values: ASCP practice parameter. Am J Clin Pathol 1997;108:247–253.

Lundberg GD. Critical (panic) value notification: an established laboratory practice policy (parameter). JAMA 1990;263:709.

Table 67.
Elements of a Critical Values Policy

- Written policy
  - Reviewed annually
  - Information distributed in training
  - Policy manual available to staff
- Identify critical tests and threshold values
- Action policy if critical value occurs
  - Verification of test results
  - Policies for notifying clinical staff
  - How and when result should be reported
  - Who should receive results
- Documentation in response to critical values

5K. Critical Values for Glucose

Very low blood glucose values are an acute emergency. Most laboratories use a value of about 40 mg/dL (2.2 mmol/L) as a low critical value, but survey of a large number of laboratories in the U.S. shows a range of values (Table 68). Newborn babies tend to have low blood glucose values until regular feeding begins, and lower critical limits are sometimes applied for a newborn nursery.

High blood glucose levels often represent poor control of diabetes and can cause diabetic ketoacidosis or hyperosmolar coma. The critical value for high blood glucose is usually set at ~450 mg/dL (25 mmol/L). For testing with glucose meters, the upper critical value may be limited by the highest value that can be measured by the meter. All meters in common use in the U.S. measure up to at least 450 mg/dL, and many newer meters have extended these ranges higher. In some outpatient settings, it is desirable to set a lower upper limit than for inpatients to assure prompt follow-up of poorly controlled diabetes. If there is not prompt attention to test results for outpatients, effective follow-up of results may be delayed until the next clinic visit.

There is no universally true set of limits for critical values, and surveys of critical limits for glucose at medical centers in the United States vary widely, especially for upper critical limits. There have been no clinical trials to scientifically establish which threshold values should be used. Limits have traditionally been based on expert medical opinion.

In this author's opinion, the critical limits for glucose should be set by the laboratory director of bedside testing in consultation with physician groups to meet specific clinical needs of particular care settings. In large hospitals or healthcare organizations, it is desirable for a medical board or the chief-of-staff's office to review critical values in order to gather broad-based input and support from physicians.

Reference

Kost GJ. Critical limits for urgent notification at US medical centers. JAMA 1990;263:704–707.

Table 68.
Critical Values for Glucose at U.S. Medical Centers

	Average Value	Range of Values
Low limit	< 46 mg/dL	30–70 mg/dL
(newborns)	< 32 mg/dL	20–50 mg/dL
High limit	>484 mg/dL	110–1000 mg/dL
(newborns)	>326 mg/dL	200–700 mg/dL

Values from a national survey of US medical centers

Reference

Emancipator K. Critical values: ASCP practice parameter. Am J Clin Pathol 1997;108:247–253.

Regulatory Issues

6A. CLIA '88 and Regulations Applying to Bedside Testing

The Clinical Laboratory Improvement Act of 1988 (CLIA '88) was a response to reports of inaccurate laboratory testing described in the Wall Street Journal and other news publications. The most severe problems reported involved Pap testing for cervical cancer, but the resulting laws were applied broadly to all laboratory testing, including physicians' office laboratories and bedside testing in healthcare organizations. CLIA defined laboratory testing as any analysis of specimens collected and removed from the body, such as collection of blood for glucose testing. CLIA does not regulate home testing, research testing not used for medical treatment, or forensic testing. Also, if measurements of glucose or other parameters are performed directly by sensors attached to or within the body, such as by cardiac monitors, oximeters, or analyzers of anesthetic gases, these are generally classified as monitoring rather than as laboratory testing.

The federal law implementing CLIA '88 required several years for amendment and enactment and was published in 1992 in the Federal Register (February 1992, volume 57, pp.7001-7288). This document detailed federal regulations for the licensing of all laboratories and the standards they must meet. The Center for Disease Control and Prevention (CDC) helps set technical standards and the Health Care Financing Administration (HCFA) has responsibility for certification processes and supervision of accrediting programs, proficiency-testing programs, and laboratory inspections. A Clinical Laboratory Improvement Advisory Committee (CLIAC) serves as a forum for public input and advice about changes in regulations.

It is important to have up-to-date information about CLIA because new information and amendments to regulations occur when new test procedures are classified, new accrediting or proficiency testing organizations are accepted, or when Congress introduces changes to the law (for example, there have been several legislative attempts to decrease the regulation of physician's office laboratories).

Table 69.
CLIA '88 Regulations

- Licensing of all laboratories
  - Even waived testing requires a certificate of waiver.
  - License fees are based on test volume and scope of testing.
- Laboratory testing classified into 4 levels of complexity
  - Physician-performed microscopy
  - Waived testing: for least complex procedures and tests approved for home use such as glucose meters designed for home use
  - Moderate complexity—most routine laboratory tests, including many glucose meters for hospital use and central laboratory glucose test procedures
  - High complexity—tests considered to require special training and expertise
- Personnel standards depend on complexity of tests
- Proficiency testing required except for waived testing and physician-performed microscopy
- Laboratory inspections

References

Bachner P. Is it time to turn the page on CLIA '88? JAMA 1998;279:473–475.

Berger D, Buttitta P, Trotto NE. MLO's national CLIA '88 survey, part 2: praise vs. protest—views from both sides of the fence. MLO 1998(June);30:36–53.

Ehrmeyer SS, Laessig RH. Regulatory requirements (CLIA '88, JCAHO, CAP) for decentralized testing. Am J Clin Pathol 1995;104(Suppl 1):S40–S49.

Passey RB. A master plan for implementing CLIA. MLO 1992(November):36–41.

Baer DM. Point-of-care testing: regulatory and reimbursement issues. Clin Chem 1992;38:1138.

6B. Applying for CLIA Certification

Every organization that is performing laboratory testing, even if the tests are only simple tests such as fecal occult blood testing or checking blood glucose levels with glucose meters approved for home use, is required to apply for a certificate of waiver or a certificate of accreditation that must be renewed every two years. Licenses are required in order to receive payment for laboratory tests from Medicare, and HCFA statistics for 1996 indicate that more than 150,000 laboratories have been licensed. Slightly more than half of laboratories are in physician offices and about 40% of the laboratories have a certificate of waiver.

There is one major exception: in some states, laboratory regulations meet CLIA standards and the states are considered CLIA-exempt. The exemption covers all laboratories in Washington and Oregon, and all but physician office laboratories in New York. Other states, including California and Georgia, have applied for exemption of their programs. Laboratories in exempt states apply to their state agency for accreditation rather than to HCFA.

Not every testing site within an organization needs a separate CLIA certificate (often commonly referred to as a CLIA license). If there are 10 nursing units in a hospital performing glucose testing, all of the testing sites can be covered by a single certificate. If the testing sites are at separate addresses, for example, 10 clinics distributed around a county or city, each should have a separate certificate.

Testing performed at multiple locations—as, for example, by visiting nurses, health fairs, or ambulances—may be covered by a certificate for the home base of operations. Minimizing the number of certificates is desirable from the standpoint of reducing expenses and paperwork. It is possible to combine all testing, whether it is performed in a central laboratory, satellite laboratories, or at the patient bedside, under a single certificate. Sometimes, a central laboratory is reluctant to be included in the same certificate with bedside testing, especially if the laboratory has little input regarding testing supervision and if there is a perception that the bedside testing does not meet usual laboratory standards of practice. Combining all bedside testing, such as whole blood glucose analysis, fecal occult blood, and urine dipstick testing, into a single certificate application might be a practical option when bedside testing is not managed by the laboratory, when different accreditation standards may be applied (for example CAP for laboratory and JCAHO for bedside testing), or when the laboratory refuses to be included in the same license.

Reference

Passey RB. How to obtain a CLIA certificate without getting stuck on the details. MLO 1992;24(8):26–31.

Table 70.
CLIA Certification

- Required of all laboratories even those performing waived tests
  - Except in CLIA exempt states—apply to state agencies
    - Washington and Oregon, all laboratories
    - New York, all labs except physician office labs
    - Other states have applied.
- How many licenses are needed for bedside testing?
  - All nursing units and laboratories at a single address can be combined.
  - Keep the number of licenses to a minimum—decrease expenses and paperwork
  - Reasons for different licenses for bedside testing
    - Different department and directors—reluctance to be responsible for other departments
    - Different accrediting agency for central lab and bedside testing, e.g., CAP vs. JCAHO
    - Concern about one component failing accrediting standards
  - Testing at different addresses should have separate licenses.

6C. Deciding Type of Certification

Whether to apply for a certificate of waiver or other certification depends on what tests will be performed and what equipment will be used. Obtaining a certificate of waiver may be a significant advantage for a clinic or physician's office laboratory in terms of decreased paperwork and expense for regulatory oversight.

For glucose testing, some meters have been classified as waived testing and others have not, depending on whether they are approved for home use or whether the manufacturer has applied for classification as a waived test. If glucose testing is to be performed under a certificate of waiver, it is important to select a meter that is classified as a waived test. This information can be obtained from test vendors or from HCFA. Test classification is a matter of public record. There is less advantage for hospitals to have testing performed under a certificate of waiver, because standards of the JCAHO meet or exceed the standards of CLIA for moderate complexity testing under a certificate of accreditation.

Certification of moderate or high complexity testing is a two-stage process. After the initial application is approved, a certificate of registration is issued. Then after satisfactory completion of an inspection by a state surveyor or another CLIA-approved accrediting agency, the laboratory is granted a certificate of compliance (surveyor) or certificate of accreditation (other approved agency).

Table 71.
Which Certification?

Waived testing only

- Certificate of waiver ($150 fee as of 1/1/98)
- Eligibility depends on tests performed and equipment used: some glucose meters waived, others not
- Advantage for clinics and physician office laboratories: decreased cost and regulatory oversight
- Less advantage for hospitals: standards of JCAHO must be met and these exceed CLIA standards

Moderate or high-complexity testing—applies to bedside glucose testing if meters are not classified as waivered. Two steps to certification:

1. Certificate of registration (fee $100) issued after completed application but before inspection
2. Certificate of compliance (fee $150-$7,940 based on test volume) issued after inspection by a state surveyor

   OR

   Certificate of accreditation (fee $150-$7,940) issued after inspection by a CLIA-approved agency other than a state surveyor

6D. Decreased Regulations Under a Certificate of Waiver

There are several major differences in federal regulations for laboratories that have a certificate of waiver, and the application process is less complicated and expensive. To be eligible, laboratories must perform only the limited menu of procedures that have been classified in the waived category. For glucose testing, meters that have been approved for home use are classified as waived. Some meters designed for hospital use have not been classified as waived. If glucose testing is performed under a certificate of waiver, use a waived test device. Manufacturers' instructions for performing tests must be followed.

Laboratories operating under a certificate of waiver are exempt from the personnel standards and requirements for quality assurance, including proficiency testing and quality control. There are no routine scheduled inspections, but a small number of laboratories are inspected on a random basis or in response to complaints.

However, good standards of practice for glucose testing should lead any laboratory to perform routine quality control testing and to participate in some form of proficiency testing. In a hospital setting accredited by the JCAHO, the required standards meet or exceed standards required for moderate complexity testing under CLIA.

Table 72.
Advantages of a Certificate of Waiver

- Simpler application
- Lower fee
- Exempt from requirements
  - Personnel standards
  - Proficiency testing
  - Quality control
  - Inspections every two years
- Regulations that do apply
  - Follow manufacturers' instructions for tests
  - Cooperate with random inspection
  - Notify HCFA within 30 days of changes in ownership, director, address, or name

6E. CLIA Certification Requirements

Applications for licenses are processed by the Health Care Financing Administration[1]. Fees are based on the type of license and the volume of testing; fee collection was mandated so the program would be self-supporting. Renewal is required every two years. The application must include the organization's Medicare/Medicaid provider number and identify personnel who will have specific roles within the laboratory, such as serving as the laboratory director. Laboratory directors are limited to overseeing laboratories covered by a maximum of five CLIA licenses. There are specific personnel requirements for directing operations and for performing tests. The minimum federal standard for personnel performing tests, unless the test is of high complexity, is to have a high school degree or equivalent and to have appropriate training with the test procedure.

Once licensed, laboratories undergoing changes such as change of ownership, facility name, address, or director must notify HCFA within 30 days. When there are changes in test menu or test methodologies, labs must notify HCFA within 180 days (six months). Inspections must be allowed even if there is no prior notice, and access to records must be provided. Proficiency testing is required, except with a certificate of waiver, and it must be performed by the staff that routinely perform the test and in the same manner as a patient sample. Manufacturers' directions must be followed and reagents used within stated expiration dates. If procedures are modified by a laboratory, additional validation of performance must be completed, and the test procedure is considered to be of a higher complexity. Performance of all quantitative tests should be verified before use, and at least two levels of quality control testing should be done every day for quantitative tests such as glucose. Some tests, as indicated by regulations or manufacturers' directions, require more frequent control testing. When multiple analyzers or methods are used to measure the same analyte, there should be a periodic comparison of the performance of all analyzers. (The original standard was to perform this comparison twice every year.)

Reference

Passey RB. A master plan for implementing CLIA. MLO 1992;24(9):36–41.

---

[1]Health Care Financing Administration, Division of Outcomes and Improvements, HCFA, 7500 Security BLVD, Baltimore MD 21244; 410-786-3531.

Table 73.
CLIA Certification Requirements

- License Fee
  - Determined by size of test menu and test volume
- Application
  - Renew every 2 years
  - Identify personnel, especially director
  - Medicare/Medicaid provider number
  - Test menu and approximate volume of testing
- After receiving certificate
  - Notify HCFA about changes within 30 days:
    - Owner, organization name, address, director
  - Notify HCFA about changes within 6 months
    - Changes in test menu and major methodology changes
  - Inspections of laboratory
    - No advance notice required
    - Records must be provided
  - Detailed policies and procedures for tests
  - Follow manufacturers' directions for tests: modifications require validation and higher complexity level.
- Quality assurance
  - Evaluate total testing process.
  - Have a process for identifying and correcting problems.
  - Perform proficiency testing.
  - Implement quality control: check at least two levels daily for quantitative tests.
  - Validate performance of tests.
    - Accuracy, precision, lowest detection limit (sensitivity)
    - Effect of interferences (specificity), reference or normal range, usable measuring range
  - Periodic comparison of multiple analyzers and methods
- Documentation—retain all records for a minimum of 2 years.

6F. CLIA Personnel Requirements

CLIA defined minimum personnel standards for performing tests and supervising and directing laboratory testing of moderate and high complexity. It is important for each laboratory to document that all staff members meet the educational and training requirements for their specified roles. In some cases, state regulations or institutional policies define higher levels of training, experience, or specific certification to perform laboratory testing or to serve as a laboratory director. For moderate complexity testing, which includes all glucose testing except when performed by waivered procedures, testing personnel must at least have high school degree or an equivalent such as the GED and adequate training for and testing of competence. For bedside glucose testing, this means that nurses, nursing assistants, phlebotomists, or other staff meet minimal CLIA standards for testing personnel as long as they have a high school diploma or equivalent.

CLIA defined three other positions—the director, technical consultant, and clinical consultant—for testing programs. The three roles can be performed by a single person with suitable qualifications or by as many as three different individuals. The director is responsible for overall planning, quality assurance, and for specifying in writing which personnel are authorized to perform specific duties. Responsibilities of the technical consultant include providing technical support for evaluation of testing quality and setting standards for training and assuring the competency of each staff member. The director and technical consultant for moderate complexity testing must meet similar minimal requirements of a bachelor's degree and at least two years of additional training or experience. The clinical consultant is responsible for reviewing the clinical appropriateness of testing and reports and for providing clinical consultation. Nursing staff usually are not qualified to fulfill any of these three roles for moderate complexity bedside testing. Participation of a medical technologist or other trained laboratory worker with a bachelor's degree helps fulfill the technical consultant and possibly the director roles. Due to the complexity of the regulations, it is clearly beneficial to have a trained laboratorian assist with the application and supervision of processes. Often the director and medical consultant positions may be filled by the same individual (with doctoral training). Physicians fulfilling the director role should be certified in anatomic or clinical pathology or have at least 20 hours of training in directing clinical laboratories.

Reference

Passey RB. How to meet the new personnel requirements. MLO 1992;24(9):47–51.

Table 74.
Personnel Requirement for Moderate Complexity Testing

- Personnel performing tests

  Minimum of high school diploma or equivalent

  Training in test performance and check of competence

- Director

  Minimum of bachelor's degree in scientific field and two years of experience as a laboratory supervisor

  Responsibilities: review of policies, compliance with CLIA standards, procedures, quality assurance, and personnel

  Limited to directing a maximum of five laboratories

- Technical Consultant

  Bachelor's degree with a minimum of two years' experience in the appropriate laboratory specialty

  Responsibilities: technical and scientific issues of testing

- Clinical Consultant

  Physician board certified in pathology or with at least 20 hours of training in direction of laboratories or Ph. D. with certification in a laboratory discipline

  Responsibilities: review of clinical appropriateness of policies and clinical consultation

6G. CLIA Inspections

Laboratories certified to perform moderate or high-complexity testing are inspected on-site for initial certification and then every two years. Inspections can be either through an accrediting agency such as CAP, which has deemed authority from HCFA, or by state surveyors who act as agents for HCFA. In most states, surveyors are members of departments of health or public health. HCFA has provided standards and interpretative guidelines for inspectors to follow.[2] State inspectors generally will have training as medical technologists and work full-time dealing with certification issues and inspections. Inspections should include direct observation of the complete testing process from sample collection to reporting, interview of personnel, and review of records of personnel qualifications, proficiency testing, and quality assurance activities. Documentation of all necessary activities and retention of records for a minimum of two years is required. There may be additional fees for the participation in accreditation programs that perform inspections and for the time spent by state inspectors. Check with state or accrediting agencies to obtain detailed information about the cost of inspections.

If a laboratory has established a good record of performance, it may be possible for it to reduce on-site inspections to a frequency of once every four years. The survey that would occur after two years may be replaced by completing and mailing an Alternate Quality Assessment Survey.

Laboratories operating under a certificate of waiver have no scheduled inspections. However, a small proportion of laboratories are randomly selected for inspection. Also, inspections could occur in response to complaints to HCFA. Inspectors would seek to verify that the laboratory is performing only waived tests and that the tests are performed according to manufacturers' directions.

Reference

Jahn M. CLIA, part 2: dealing with inspections and paperwork under CLIA. MLO 1994;26(5):30–36.

---

[2]Standards and guidelines are available in publication PB92-146174 from the National Technical Information Service, 5285 Port Royal Road, Springfield VA 22161; 800-553-6847.

Table 75.
CLIA Inspections

- Frequency of on-site inspections
  - For initial certification
  - Then at 2-year intervals
  - With good performance
    - Alternative quality assessment survey at 2 years
    - On-site inspection at 4 years
- Inspectors
  - Employees of state agency
    - Full-time review certification and perform inspections
    - Usually trained as medical technologists
  - Accrediting agency
    - Examples—JCAHO, CAP, COLA
    - Must meet or exceed HCFA standards
- The inspection process
  - Guidelines developed by HCFA
  - Elements of inspections
    - Direct inspection of complete testing process
    - Interview of employees
    - Review of records
      - Test menu
      - Personnel qualifications
      - Training
      - Proficiency testing
      - Quality control

6H. Penalties For Not Following CLIA Rules

Failure to comply with CLIA regulations has several potentially serious consequences. The potential severity of penalties explains why laboratories and laboratory directors do not want responsibility for testing operations when adequate mechanisms or authority for meeting regulatory requirements is absent. This can make them reluctant to include bedside testing in their areas of responsibility.

The primary penalty for rules violation is suspending, limiting, or revoking laboratory certification. This renders the laboratory ineligible for receiving Medicare and Medicaid payments for laboratory testing. Also, anyone who owns or operates a laboratory that has had its certification revoked may be barred from owning or operating a laboratory for two years. Other penalties that may be imposed include a directed corrective action plan, on-site monitoring by surveyors, and fines of up to $10,000 per day or per violation. Penalties result in inclusion in an annual registry of adverse actions against clinical laboratories by the federal government If a laboratory is considered to represent a risk to patients, testing may be halted.

Some examples of actions that may result in sanctions include refusal to allow an inspection or access to records, performing tests not covered by the laboratory's certification, lying on applications for certification, reporting results for tests that were not performed, not complying with regulations identified during inspections, excessively failing proficiency testing, or performing proficiency testing in a special manner that is different than that used for patient testing.

Table 76.
Penalties For Not Following CLIA Rules

- Suspension or loss of certification
  A consequence is loss of ability to bill Medicare or Medicaid for lab tests
- Barring from ownership or operation of any laboratory for two years
- Fines up to $10,000 per day or per violation
- On-site monitoring
- Corrective action plans
- Halting of testing activities

6I. Accrediting Agencies for Laboratories

HCFA has reviewed the standards of a number of organizations that perform laboratory accreditation, and it has concluded that they meet or exceed CLIA standards. These organizations have "deemed status" for CLIA. Some of these organizations, such as the American Society of Histocompatibility and Immunogenetics and the American Association of Blood Banks, are approved only for limited laboratory specialties. The organizations that are likely to provide accreditation for programs including bedside glucose testing are the Joint Commission on Accreditation of Healthcare Organizations (JCAHO), College of American Pathologists (CAP), American Osteopathic Association (AOA), and the Commission on Laboratory Accreditation (COLA). Most hospitals undergo JCAHO inspection of their clinical operations. However, JCAHO recognizes laboratory inspections by CAP and the state of Washington so that laboratory inspection would not have to be repeated if the laboratories are accredited by those agencies.

Most often, accreditation of laboratories performing bedside testing in hospitals is sought from the JCAHO or CAP. The AOA has provided accreditation in a smaller number of hospitals. COLA was established as a program for accreditation of physician office laboratories.

Table 77.
Organizations Accrediting Laboratories for CLIA

American Osteopathic Association (AOA)
142 East Ontario
Chicago, IL 60611
312-280-5898

College of American Pathologists (CAP)
325 Waukegan Road
Northbrook, IL 60093-2750
800-323-4040
www.cap.org

Commission on Laboratory Accreditation (COLA)
PO Box 247
Columbia, MD 21045-0247
800-298-8044
www.cola.org

Joint Commission on Accreditation of Healthcare Organizations (JCAHO)
Department of Laboratory Accreditation Services
One Renaissance Blvd.
Oakbrook Terrace, IL 60181
708-916-5783
www.jcaho.org

6J. Standards of the College of American Pathologists (CAP)

CAP considers that the same high standards for testing should be applied whether a test is performed in a central laboratory or at a patient's bedside. Also, CAP recognizes no difference in standards regardless of whether the test procedures are classified under CLIA regulation as being waived or as being of moderate complexity. CAP has deemed authority by CLIA (that is, CAP accreditation is accepted by HCFA), and standards generally exceed those required by CLIA.

CAP inspection standards are outlined in checklists relevant to each laboratory section. A specific checklist (#30) has been developed for "Point-of-Care Testing," which is characterized by performance of tests outside a fixed location. Small satellite or stat laboratories in intensive care units are not considered point-of-care testing, and these laboratories are evaluated with checklist 25 (Limited Service Laboratory) or 26 (Blood Gas Laboratory). Checklist items are updated annually.

Inspections emphasize having appropriate written procedures (which must be in the format recommended by the National Committee on Clinical Laboratory Standards [NCCLS]) and documenting performance. Participation in CAP proficiency-testing programs was previously required; now other proficiency-testing programs are acceptable. All laboratory operations must be reviewed with respect to the general checklist, which includes review of adequacy of facilities, record-keeping, data systems, and safety issues.

On-site inspections are performed every other year by teams of inspectors trained and assembled by the Laboratory Accreditation Program of the CAP. Laboratories receive advance notice of inspections, and inspections are usually scheduled one to two months in advance. Inspectors are usually laboratory professionals who donate their time to performing one or two inspections per year.

Inspectors usually have ongoing daily experience in the laboratory areas that they inspect. They offer the opportunity to discuss potential deficiencies that are identified during the inspection. The checklists contain a large number of questions that are classified as Phase 0 for informational items and as Phase I for issues that are important for an outstanding laboratory. Additional questions, classified as Phase II, pertain to issues considered essential for accreditation of the laboratory. Each checklist item is evaluated as "pass" or "fail." At the end of the inspection, a report of deficiencies and recommendations for improvement is provided. The laboratory must respond in writing to deficiencies, and deficiencies with respect to Phase II questions must be corrected.

In alternate years a laboratory must perform a self-inspection based on the point-of-care checklist. CAP considers the inspection process both a means of meeting regulatory standards and an educational process. Inspectors are encouraged to offer suggestions for improvement.

Reference

Ehrmeyer SS, Laessig RH. Regulatory requirements (CLIA '88, JCAHO, CAP) for decentralized testing. Am J Clin Pathol 1995;104(Suppl 1):S40–S49.

Table 78.
CAP Inspection of Bedside Testing

- Inspection every two years by CAP inspectors
  - Prior notification and scheduling of inspections
  - Alternate year self-inspection
- Inspectors are volunteers who are laboratory workers.
- Standards are in checklist 30 "Point-of-care testing."
  - There are three categories of questions:
    - Phase 0: informational
    - Phase I: important for optimal performance
    - Phase II: required for accreditation
  - Scoring is pass or fail.
- Specific format for procedure manual (NCCLS)
- Deficiencies
  - Written report at time of inspection
  - Phase II deficiencies must be corrected.
    - Written response to CAP

Reference

National Committee on Clinical Laboratory Standards. Clinical laboratory procedure manual. Approved Guideline (GP2-A2). Villanova PA: NCCLS, 1992.

6K. Preparing for a CAP Inspection

Begin preparing a year or more in advance of an inspection to allow enough time for experience with proficiency testing and for documenting all testing procedures. After enrolling in a proficiency-testing program, for example, several months may pass before the order is processed and the first samples for testing arrive; after sending in results, one or two months may pass before summary reports are obtained. Creating detailed procedure and policy manuals also can consume months of work. The CAP directs that all procedures should be in a specific format prescribed by NCCLS. These need to be reviewed and approved by the director annually, or whenever a new director is assigned for testing. Training is a lengthy process when there are large numbers of testing personnel, and it is an ongoing process because there must be a procedure for periodic checking and documentation of personnel competence. Demonstration of adequate performance of quality control programs requires at least a couple of months of experience.

To prepare for the inspection, the testing program should be evaluated against each checklist question. Note what documentation is available to support it. Inspections usually emphasize examples of corrective actions that have been taken in response to proficiency-testing or quality control problems. It is useful to assemble examples of these efforts. Central record-keeping greatly simplifies inspections. If records are not maintained centrally, inspectors are supposed to review each location with a separate checklist, and this may increase the billing for an inspection.

Perform an internal inspection of all testing sites prior to a formal inspection to ensure that all materials are dated and stored appropriately and that expired materials are removed even if they are no longer in use.

Table 79.
Common Deficiencies Noted in CAP Inspections of Bedside Testing

- No reporting of normal ranges together with test results
- Inadequate labeling of reagents regarding date opened, received, or expiration
- Lack of criteria for verifying the reportable range of methods
- Lack of a documented system for detecting clerical errors
- Lack of daily evaluation of quality control data
- No evidence of corrective action when control results are out of range
- Lack of a document describing the evaluation of quality controls
- Lack of review of test results by a supervisor and by the next routine working shift
- Incomplete procedure manuals
- Failure of director to review procedures annually or at change of director
- Lack of weekly review of quality control values
- Lack of documentation of periodic competency checks of all personnel
- Expired reagents at testing sites

6L. JCAHO Standards

The JCAHO, like CAP, does not recognize substantial differences between performance standards for waived testing vs. moderate complexity testing as classified by CLIA '88. One slight accommodation noted by the author during a previous inspection was allowing the manufacturer-determined range for quality control testing of glucose meters rather than requiring establishment of ranges specific for the laboratory. Accreditation manuals describe standards for the organization and for laboratory testing. Manuals are updated periodically so it is important to have the current version to verify that your institution is in compliance.

There have been significant differences between standards and surveys of the JCAHO compared to CAP. JCAHO surveys have used a quantitative rather than a pass-fail approach for scoring. Scoring on a specific issue such as quality control testing will depend on the compliance rate and on how long an adequate program has been in place. Meeting standards one week before the inspection occurs would not necessarily be reflected in a bad evaluation by CAP, but would lead to poorer scores by the JCAHO. To achieve the highest scores, a program may have to be in place for one year or more; therefore, long-term preparation is important. The JCAHO uses professional surveyors rather than volunteers. JCAHO inspectors tend to spend more time observing laboratory processes and evaluating their impact on patient outcomes than do CAP inspectors. Recently, the JCAHO placed increased emphasis on performance improvement programs, which are an updated version of total quality management, and a new initiative termed Oryx aims to monitor specific performance indicators for hospitals.

Common deficiencies identified by the JCAHO include lack of corrective action for quality control or proficiency testing, lack of proficiency-testing programs for regulated analytes (glucose is a regulated analyte), and inadequate assessment of personnel competence for performing tests.

Table 80.
JCAHO Standards

- Little difference in standards for waived vs. moderate complexity testing
- Quantitative assessment rather than pass-fail
  - Scores based on level of compliance and how long programs have been compliant
- Professional surveyors
- More extensive evaluation of processes
  - More direct observation of processes
  - Evaluation of impact on outcomes
  - Evaluation of performance improvement plans

References

Belanger AC. The Joint Commission and CLIA: a 5-year retrospective. MLO 1998;30(2):46–48.

Holmes RL. Conquering performance improvement documentation for JCAHO. MLO 1998;30(6):18–24.

Joint Commission on Accreditation of Healthcare Organizations. Accreditation manual for pathology and clinical laboratory services. Oakbrook Terrace, IL: JCAHO, 1997.

Joint Commission on Accreditation of Healthcare Organizations. Comprehensive accreditation manual for hospitals. Oakbrook Terrace, IL: JCAHO, 1995.

Belanger AC. Point-of-care testing: the JCAHO perspective. MLO 1994;26:46–49.